SPITFIRE!
SPITFIRE!

GW00363007

Michael Burns

SPITFIRE! SPITFIRE!

BLANDFORD PRESS
POOLE · NEW YORK · SYDNEY

First published in the UK 1986
by Blandford Press Ltd,
Link House, West Street,
Poole, Dorset, BH15 1LL

Distributed in the United States by
Sterling Publishing Co., Inc.,
2 Park Avenue, New York, NY 10016

Distributed in Australia by
Capricorn Link (Australia) Pty Ltd
P O Box 665, Lane Cove, NSW 2066

ISBN 0 7137 1832 3

British Library Cataloguing in Publication Data

Burns, Michael G.
Spitfire! Spitfire!
1. Spitfire (Fighter planes)
I. Title
623.74'64 UG1242.F5

Printed in Great Britain by
R. J. Acford Ltd., Chichester, Sussex.

Title Page
Spitfire Mk VIII JF880 AN-U of No. 417 (City of Windsor) Sqn, Royal Canadian Air Force, at Al Fano, Italy, on 29 December 1944.

Contents Page
Spitfire F Mk XIVE RB145 built by Vickers-Armstrong (Supermarine).

Acknowledgements

The author gratefully acknowledges the particular assistance in the preparation of this book of John Batchelor, Cynthia Brown, Gilbert F. Burns, Julia Burns, Group Captain H. C. Daish, Lewis Deal, John Douet for his excellent design, the Christopher Elliot Collection, John Fawcett, Jeremy Flack, Don Healey, John Hibbert, Ian Huntley, David Innes, Robert Jackson, the Kent County Council, *The Kent Messenger*, Alec Lumsden, Michael O'Leary, John Newth, Bruce Robertson, Don Robertson, The Smithsonian Institute, Geoffrey Thomas, *The Times* and Fred W. Town.

Contents

R. J. Mitchell

The first Spitfire was a failure. Reginald Mitchell designed it to meet the Air Ministry's demands for a fighter that was truly modern, expressed in Specification F7/30, issued in autumn 1931. By 1930, the twin-gun, open-cockpit, fabric-covered fighter biplane had almost reached a plateau of perfection, epitomised in the RAF's Bulldog and Fury, by slowly refining rather than using new approaches. In contrast, civil and racing designs had advanced in sophistication and performance by leaps through the 1920s. It was time for a new departure in fighter design.

This was precisely the kind of challenge upon which Reginald Mitchell thrived. He was not only new to the fighter field, but an innovator; he was also a practical man who knew that even the soundest technical knowledge without common sense resulted in mediocrity.

Reginald Joseph Mitchell had been born in Talke village, near Stoke-on-Trent on 20 May 1895. Leaving school in 1911, aged 16, Reginald joined the famous locomotive engineers Kerr Stewart & Co., Stoke. Completing his apprenticeship, he began working in the drawing office. At night school, he studied engineering, mechanics and higher mathematics and, at home, mastered practical engineering on a lathe.

The Great War accelerated aviation's development and aircraft manufacturers thrived upon Government orders. Since boyhood, Mitchell had been keenly interested in aviation. He devoted his spare time to building and flying model aircraft, but wanted to be more closely involved. In early 1917, aged 21, he joined the Supermarine Aviation Works Ltd., as a draughtsman.

When the war ended in November 1918, the Government's large aircraft production orders were cut. Many companies foundered but, astutely managed by Hubert Scott-Paine, who appointed Mitchell chief designer, Supermarine's survival turned into success and then expansion.

During the 1920s, Mitchell pursued three main categories of aircraft: single engine amphibians from 1919; twin-engined flying boats from 1925; and racing floatplanes from 1925. The series of amphibians pioneered commercial air services and founded Supermarine's reputation, but included three racing flying boats.

As seaplane manufacturers, Supermarine were attracted by the Schneider Trophy contests, but, as the Government did not support teams, had to enter privately. For the 1919 contest at Bournemouth, Hargreaves modified the Baby, a wartime amphibian design, to take a 450 hp Lion. It crashed before the start; indeed, no one completed the course. Italy, unopposed, won in 1920 and 1921; a win in 1922 would secure Italy the Trophy outright. Mitchell redesigned the Sea King II — the first design over which he had control, an amphibian fighter which did not enter production — with a 450 hp Lion, as the Sea Lion II. The only challenger to the Italians, flown by Captain Henri C. Biard, it won and, to cap the triumph, was also granted four Marine World Records.

The graceful Southampton first marked Mitchell as a great designer. A twin-engined, long-range, five-crew reconnaissance flying boat, the Air Ministry ordered six in August 1924, uniquely, from the drawing board. Embodying all Mitchell's experience, the Southampton-series set new standards. The Mark I was of wood construction, the Mark II had a metal hull while the Mark III had an all-metal airframe, demanded for RAF aircraft from 1925,

and, ultimately, Mitchell developed it into the successful Scapa and Stranraer in the 1930s. Such comprehensive long-term development set Mitchell apart as a designer.

In 1928, Vickers (Aviation) Ltd acquired the share capital. A condition of purchase was that Mitchell would bind himself to the company until December 1933. With a place on the Board of Directors, he agreed. To modernise production methods, Vickers' chairman appointed the forceful Trevor Westbrook as General Manager. Backed by the resources and political power of Vickers, Supermarine's prospects were excellent.

The American victory in the 1923 Schneider Contest, on their first attempt, was a turning point. Speed now dominated performance; flying boats could not compete against floatplanes. To beat the American floatplanes, Mitchell needed a powerful engine in a light airframe with low frontal area. This time, the Air Ministry provided funds. Unavoidably, flying boats have large frontal areas and high drag, and Mitchell abandoned an amphibian of advanced design he was working on for the 1924 Contest when the Americans postponed the event until 1925.

Despite its failure, Mitchell's S.4 (S: Schneider) racing floatplane marked the beginning of the line of development that most remarkably displayed his full calibre. The 700 hp Lion-powered S.4 was the simplest and most efficient configuration; a tractor-engined, single fin/rudder, mid-wing monoplane with twin floats. The only protrusions were the underwing radiators. It was a profound piece of engineering, but it crashed before the race when flutter — then a little understood phenomenon — developed in the wing. The Americans won the 1925 Contest.

The Air Ministry now expanded their high speed research programme to include the racers, formed a Service High Speed Flight racing team at Felixstowe and funded racers, not for the 1926 but for the 1927 race — a risk, given the Americans could win the Trophy outright, but Italy's Macchi M.39 low-wing floatplane defeated the Americans in 1926.

For the 1927 Contest, Mitchell had to reduce both drag and weight and improve water performance. The S.5 was a breakthrough. To avoid the heavy drag of conventional airstream radiators but effectively dissipate the 750 hp Lion VIIA's immense heat, smooth-skinned surface water radiators covered nearly all the wing. Though numerous, the competition was all slower and less reliable than the three S.5s and fell out of the race one by one. The S.5s could hardly lose.

After 1927, the Contest was held every two years. For his next racer, the S.6, Mitchell needed 1,500 hp, over 40 per cent increase in power. The Lion had reached its limits. Prompted by the Air Ministry, Sir Henry Royce of Rolls-Royce agreed with alacrity to develop a racing engine. From this sprang a partnership that had a profound influence on Supermarine. Astonishingly, the 'R' developed 1,900 bhp at 2,900 rpm — against the Lion VII's 875 bhp — and had an unprecedented power to weight ratio of 0.8 lb/hp. The critical importance of fuel mixture was being realised, but Rodwell Banks, 'Hot Rod', solved the 'R's problems by diluting pure benzole with 22 per cent 'light cut' leaded gasoline.

Mitchell's aeroplane had to be capable of using the power. Resembling, in appearance only, a 20 per cent upscaled S.5, the S.6 was all-metal, in line with Air Ministry demands. Rather than being in three sections, the fuselage was now a single duralumin-skinned monocoque. To dissipate the volcanic heat, surface radiators covered the entire wing and most of the float upper area.

The 1929 race was a competition between Mitchell the innovator and Mario Castoldi the developer, as France, the USA and Gloster withdrew their troublesome entries, but both M.67s retired on the second lap. Flg Off. H. R. D. Waghorn in S.6 N247 at 328.63 mph led Atcherley, at 325.54 mph, across the line, but Atcherley was disqualified. Molin at 284.52 mph squeezed the M.52 across ahead of Greig in an S.5 by 2.41 mph. Now, the High Speed Flight raised the World Speed Record thrice in three days, Orlebar fixing it at 357.7 mph on 12 September.

Britain was now poised to win the Trophy outright. The British Government refused to support a 1931 team, but in January 1931, Lady Houston, a wealthy eccentric and patriot who detested Ramsay MacDonald's Government, provided £100,000 sponsorship. The only possible contender was the S.6. Under new regulations, the trials and the race had to be held the same day, requiring greater fuel capacity, hence increasing weight. Thus, the S.6 needed a 20 per cent more powerful and therefore, a thirstier, hotter engine. Rolls-Royce got 2,300 bhp at 2,900 rpm out of the 'R' but nearly 50 per cent of the S.6B's 'wetted surface' was radiator — Mitchell called it the 'flying radiator'.

The 1931 race is legendary. Beset by technical problems, France, Italy and the USA withdrew. Unchallenged, the dangerous British racers still had to complete the course to win the Trophy outright. Flt Lt John Boothman did, at 340.08 mph in S.6B S1595.

The same day, Flt Lt George Stainforth in S.6B S1596 set a new World Speed record at 379.05 mph. Even that was not enough for Reginald Mitchell, Sir Henry Royce or Rolls-Royce's experimental manager, Ernest Hives. Before the pilots, skills and organisation of the High Speed Flight were lost, they wanted to fix the Record at 400 mph. Rodwell Banks prescribed a special fuel mixture and on 29 September 1931, Stainforth in S1595 achieved 407.5 mph.

Reginald J. Mitchell.

S.6B S1595, which won the 12th and last Schneider Contest at Spithead on 13 September 1931, is now in the Science Museum, London.

Towards K5054

The Air Ministry now invited Mitchell, now indisputably a great designer, to tender for F7/30. What, on analysis, the detail requirements demanded was the best of biplane performance, merged with the best of monoplane performance, an impossible request. The eventual winner was a biplane little better than those in service, the SS.37 Gladiator.

Mitchell evaluated many configurations in Vickers' Weybridge wind-tunnel before submitting the Type 224 design to the Air Ministry on 20 February 1932. However, he modified it several times until it received 'Instruction to Proceed' in September 1933. It was powered by a 660 hp Rolls-Royce Goshawk with an evaporative steam cooling system.

A low cranked wing monoplane with fixed undercarriage of all-metal construction, it had an open cockpit and two 0.303in. guns in the forward fuselage sides and two in the wing roots, synchronised to fire through the propeller arc. The engine evaporative-cooling tanks were in the wing leading edge, with draggy corrugated skins. It lacked the finesse of Mitchell's Schneider racers and the efficient engineering of his flying boats.

Mitchell, however, was now a sick man. Late in 1933, he underwent an operation to remove an abdominal cancer. He almost died. He was told that, if there were no recurrence within five years, he would live. Much weakened, he never recovered his vitality. In 1934, he went on a continental holiday to convalesce. In Germany, he realised the political and military significance of Hitler's rise to power the previous year. He had heard of, and now saw the re-emergence of German military aviation. He resolved to design the fighter *and* the bomber that the RAF would need. He literally devoted his life to it.

On 19 February 1934, the Type 224 flew for the first time in 'Mutt' Summers' capable hands. Its performance was lower than predicted and the cooling system regularly malfunctioned. Unofficially, the Type 224 was referred to as the Spitfire. During the Type 224's Service trials Mitchell's thinking had clearly evolved beyond it. He was already experimenting with the design. In July 1934, he submitted Supermarine Specification 425a, based on F7/30.

The bombers that the Air Ministry had sought to counter with F7/30 were French. However, in June 1934, the new RAF Chief of Staff, Air Chief Marshal Sir Edward Ellington, forced upon the RAF awareness of the real implications of German re-armament: the resurgent *Luftwaffe* could soon equal the RAF in strength. In July, the Government adopted Expansion Scheme A for 84 home-based squadrons, including 28 fighter, by 1938. However, in summer 1934, the German air ministry issued a specification for a single-seat interceptor monoplane with advanced performance.

Mitchell, in his Specification 425a had recognised that what the RAF needed was a fighter able to carry a multi-gun armament to altitude, fast, to intercept incoming enemy aeroplanes flying fast and high and giving short warning of their approach. A very high performance in both climb to altitude and speed and high manoeuvreability was necessary. The design was so radically changed from the F7/30 that it was redesignated Type 300, but its predicted performance was not sufficiently better than the other designs, and the Air Ministry turned it down.

Mitchell persevered. By early autumn, the paper-fighter had thinner wings with 2ft. less span, now 37ft. 1in., a stressed skin construction and a faired cockpit enclosed fully by a Perspex canopy. The elliptical wing developed as a simple,

The French Grey prototype Spitfire, K5054, showing modified rudder horn balance.

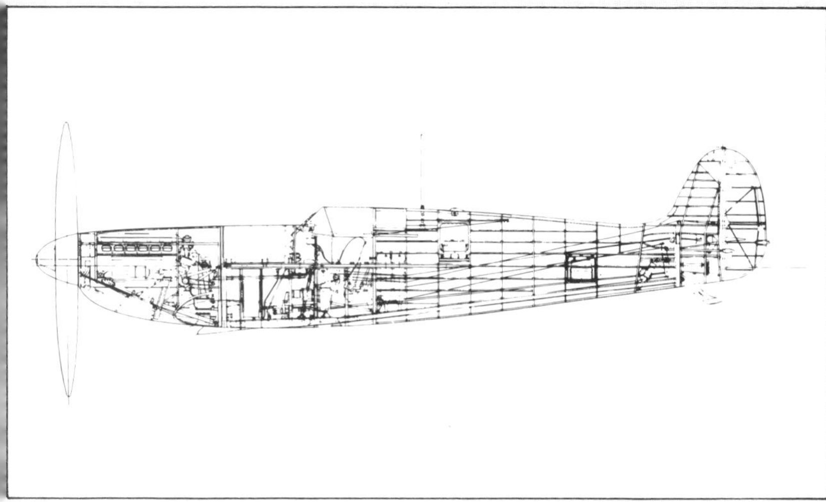

Fuselage sectional elevation of the original production Type 300 Spitfire, dated 12 June 1936. It differed in detail from K5054.

logical progression. Its predicted speed was 280 mph. It still had the Goshawk, but with a better engine, even higher speed could be achieved. The Air Ministry suggested the 700 hp Napier Dagger. This was never satisfactory. The Vickers board rejected it in favour of a new 27-litre Rolls-Royce PV-12 — the Merlin — at the suggestion of Sir Robert Maclean. Detail design work upon the PV-12/Type 300 combination was authorised immediately.

The Air Ministry, now impressed, issued a contract on 1 December 1934, for an improved 'F7/30' prototype. The contract was formalised on 3 January 1935. A new official Specification was written around Mitchell's design, F37/34, as a short addendum to F7/30.

At the same time, Sydney Camm, Hawker's chief designer, had replaced the Goshawk with a PV-12 in his 'Fury Monoplane'. It was renamed the 'Monoplane Interceptor'. The Air Ministry issued a contract and specification, F36/34.

European events formed a blackening background to all these efforts. In March 1935, Hitler boasted to British emissaries that the *Luftwaffe* already equalled the RAF's front-line strength. He exaggerated, but the British Government found it a threatening enough picture, and authorised Expansion Scheme C in May 1935, calling for 68 bomber and 35 fighter squadrons by March 1937, emphasising the introduction of modern types. In Germany, the first two monoplane fighter prototypes, the Messerschmitt Bf 109 and Heinkel He 112, were under construction.

Sqn. Ldr. Ralph Sorley at the Air Ministry, impressed by AAEE armament experiments, successfully — despite low seniority — urged the adoption of an eight-gun armament in fighters. The main 1934 fighter specification F5/34, called for an eight-gun fighter, as did F10/35. The American Colt .3in. calibre machine gun, adapted to take British rimless 0.303in. ammunition and licence-built by Browning, was selected. Mitchell accepted this requirement on 29 April 1935.

A 1/24th scale model was tested in RAE Farnborough's wind-tunnel in June 1935 to determine its spinning character-

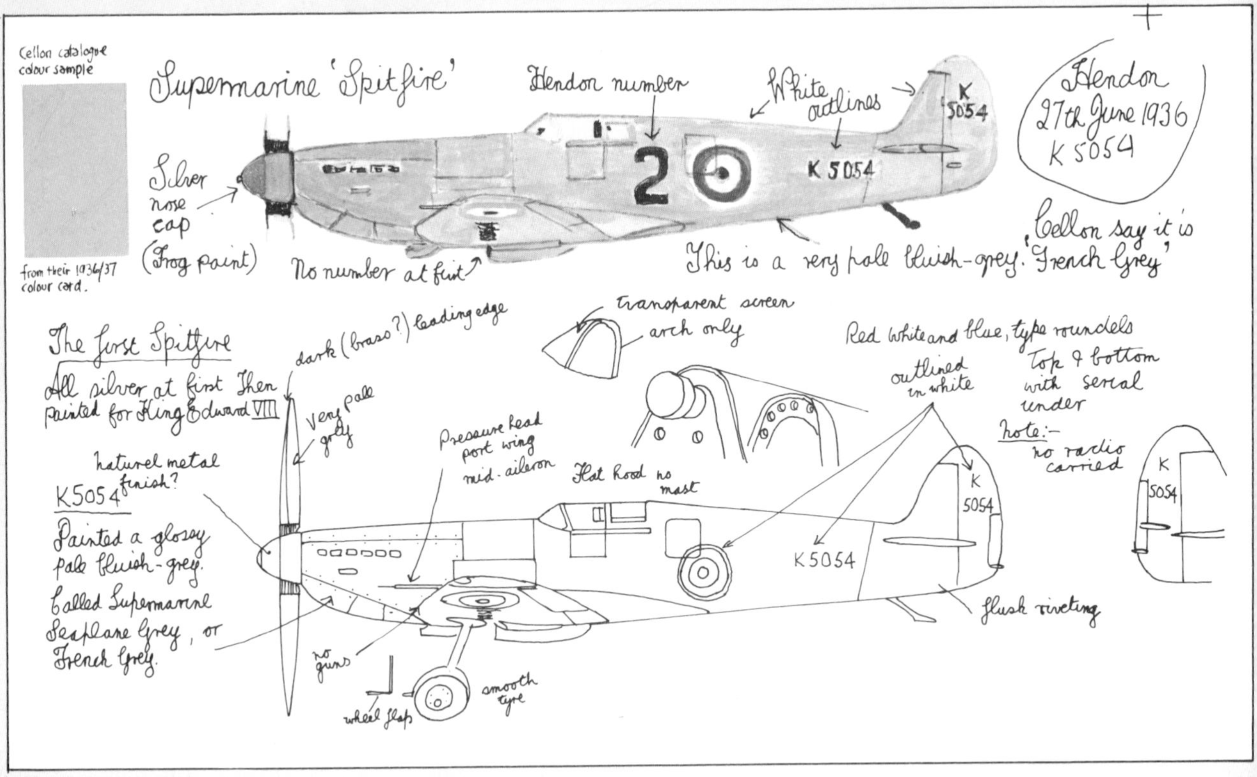

Cellon catalogue colour sample
from their 1936/37 colour card.
Supermarine 'Spitfire'
Hendon number
White outlines
K 5054
2
K 5054
Hendon 27th June 1936 K5054
Silver nose cap (Frog paint)
No number at first
This is a very pale bluish-grey. 'Cellon say it is 'French Grey''
The first Spitfire
All silver at first Then painted for King Edward VIII
dark (brass?) leading edge
very pale grey
transparent screen arch only
Red White and blue, type roundels
outlined in white
Top & bottom with serial under
Note:- no radio carried
natural metal finish?
K5054
Painted a glossy pale bluish-grey. Called Supermarine Seaplane Grey, or French Grey.
Pressure head port wing mid-aileron
Flat hood no mast
K 5054
K 5054
flush riveting
no guns
wheel flap
smooth tyre

Top: Sketch made by Ian Huntley at Hendon, with the paint chip.
Bottom: Later detail sketch.

istics. Raising the tailplane 1ft. markedly improved test results, as it no longer masked the rudder. Mitchell decided to raise it 7in. and extend the fuselage 9in.

Mitchell had sound experience of evaporative cooling systems and did not want to use the high-drag external air-cooled radiators. However, Fred Meredith of RAE had designed a new, ducted radiator which produced some thrust by expelling the air at greater velocity than it took it in, offsetting some drag. Mitchell enthusiastically adopted it. Rolls-Royce redesigned the cooling system to use ethylene glycol. This was the final major change. By August 1935, the first metal was being cut at Woolston for the prototype.

In March 1936, Expansion Scheme F was authorised, calling for 30 fighter squadrons by 1939. The Air Ministry wanted 600 Hurricanes and 300 Spitfires by this date and had been assured by Supermarine on 21 February that, given completion of Service testing at the end of April 1936, a production order on 1 May and the commencement of production in September 1937, at five a week,

K5054 at Hendon on 27 June 1936.

360-380 Spitfires could be delivered by 31 March 1939, allowing 12 squadrons to re-equip.

By early March 1936, the prototype, K5054, had completed engine running tests and the Aeronautical Inspection Directorate's airworthiness inspection. For the first flight, it was unpainted, the undercarriage was locked down and a fine-pitch propeller was fitted.

Late in the afternoon of 5 March 1936, with clear skies over Eastleigh aerodrome, fair visibility and a light west to south-west wind, Captain J. 'Mutt' Summers, Vickers Chief Test Pilot, stepped on to the wing, swung into the cockpit and slid down into the seat. Strapping himself in, he quickly ran through the engine starting routine. The Merlin C coughed into life. As the exhaust clouds cleared, he ran through the systems checks. Satisfied, he waved away the chocks. His left hand moved the throttle forward. The fighter began to roll. He taxied around, getting to know it with life in it, then, on the far side of the airfield, turned into wind, opened the throttle — and was airborne.

At 5,000ft. over open country, he tested stalling characteristics, to ensure he knew how to land, by doing a few dummy landings. Then he did a few steep banked turns to test the controls and responses. Satisfied, he opened up and headed for Eastleigh. The first flight took just 20 minutes.

He taxied to the hangar and shut down. Onlookers gathered at the port wing root. Summers scanned the cockpit, unstrapping himself and tersely conveyed that he had found no problems — so far: 'I don't want anything touched.' Everyone was elated.

The problems encountered during test flying were not major. However, the limiting speed was found to be only 380 mph (465 mph true airspeed); beyond that, wing flutter developed. Removing the rudder horn balance corrected minor directional instability. By mid-May, there was considerable pressure to get K5054 to the Aeroplane and Armament Experimental Establishment, Martlesham Heath, for official trials because the Hawker F36/34 was already there, but K5054 had not yet achieved the speed expected. With a more efficient propeller, she achieved 349 mph at 16,800ft. and a climb to 15,000ft. in 5 minutes 52 seconds.

On 26 May, Summers delivered K5054 to AAEE where she underwent an unusually rapid trials, treated as something special from the start. On 3 June, before the trials were fully complete, the Air Ministry signed a contract for 310 Spitfires.

The Spitfire and Hurricane prototypes were first shown to the public at Hendon on 26 June 1936, and there was wide public awareness of the new 'wonder machines'. This account by Ian Huntley, noted for his meticulous research into aircraft colours, camouflage and markings, conveys the mood of young boys — he was ten — anticipating seeing these machines.

It was announced in the Press as the 'Supermarine high-speed monoplane'. A real, all-metal, single-seat, world-beating interceptor fighter, we all thought — the Supermarine F37/34 was tops! It was obviously in all silver, like the earlier Hawker F36/34, and just what would it look like in colourful bands or checks with squadron leader's markings on the fin like the biplanes?

A few weeks later at Hendon, K5054 was seen finished in a gloomy greyish-blue paint scheme. 'It's because King Edward VIII is going to inspect it', they said. At that time, I could not make a match for that subtle colour but later on an uncle sent me a chipping from a Cellon paint catalogue from the SBAC Exhibition of 1936. That was excitement indeed!

Then came the depressing shock that all Expansion Programme types would be in *camouflage*! I hated the idea, but finding it to be a fact of the future, started to gather the new information as soon as it appeared. Black and white pictures showed K5054 in this dreadful scheme, which did nothing for its lines at all. I almost lost interest in RAF aircraft, but that fatal curiosity survived.

By April 1937, the headlines announced CAMOUFLAGE FOR ALL NEW RAF AIRCRAFT, and, with evidence gradually being seen with other aircraft types, a watch was kept for the Spitfire. Then, on a Saturday morning towards the end of that month, either the 24th or 31st, and about a month before the Empire Air Day on 29 May 1937, a number of camouflaged aircraft flew into Northolt. One was K5054. I noted that it was 'green and brown with silver undersides with beautiful contrasting red, white, blue and yellow roundels, but what were the other small markings . . .!

Although clearly not well, Mitchell's illness had been kept private. Resisting medical advice, he drove himself hard. He was already working on other designs, including the Type 317 long-range, four-engine bomber, the Spitfire's alter-ego, to Specification B12/36. Under the strain of designing the Types 300 and 317, his health deteriorated steadily. His wife took him to a specialist clinic in March 1937, but it was too late. Returning to Southampton, he meticulously ordered his affairs and left next month for the American Foundation in Vienna, but returned in early May. On 11 June 1937, shortly after his 42nd birthday, Reginald Mitchell died.

The unpainted K5054 (Merlin C), seen at Eastleigh just before its first flight.

Into Service

For over two years, there was just one Spitfire flying, the prototype. Spitfire production began in March 1937. Supermarine, lacking mass-production capacity, built the fuselages and subcontracted the other components to a considerable degree. Final assembly was carried out at Eastleigh airport. However, the Spitfire was complex to build and production was initially very slow.

By early 1937, the Air Ministry was concerned by the *Luftwaffe*'s rapid expansion and the threat to British interests in the Far East posed by Japanese expansionism. Scheme K now emphasised fighters. In March 1938, the Germans annexed Austria. The British Government reacted swiftly: Expansion Scheme L called for 12,000 aircraft by 1940.

On 4 August 1938, the first squadron delivery was made to No. 19 Squadron, a Gauntlet unit at RAF Duxford. It was K9789, the third production Spitfire. It received its second, K9790, on 11 August. The first and second production aircraft had been retained for thorough handling trials, and No. 19 Squadron's first two aircraft were flown intensively by Nos. 19 and 66 Squadron pilots to build up 300 hours quickly to reveal service defects. By 19 December, with gradual deliveries, No. 19 Squadron was fully equipped with Spitfires. However, by the end of 1938, only 45 Spitfires had been delivered to the RAF. The initial production contract, for 300, due for completion in March 1939, began to slip badly.

In March 1939, despite Hitler's avowals at Munich only six months before that he had no more territorial claims, Germany invaded Czechoslovakia; next month, Italy invaded Albania. Abandoning appeasement, French and British Chiefs of Staff met to decide a common defence policy against Germany and Italy. Poland, fearful that she would be invaded next, with good reason, was given a Guarantee that if Germany did, France and Britain would declare war upon Germany. A British Expeditionary Force with an RAF Air Component and an independent RAF Advanced Air Striking Force (AASF) would then be deployed to France.

The Expansion Programme, with the time won by the Munich Appeasement, was now in high gear. On 12 April 1938, the Nuffield Organisation had been contracted to build a further 1,000 Spitfires at a new 'shadow' factory, at Castle Bromwich, near Birmingham, by 1940. In 1939, on 29 April and 9 August, contracts were placed with Vickers-Supermarine for, respectively, 200 and 450. Orders totalled 2,160. At the end of August, the initial production contract was completed.

As production geared up, other squadrons re-equipped. By late 1938, No. 66 Squadron had fully re-equipped. By August 1939, there were seven squadrons within Fighter Command, Nos. 19, 41, 54, 65, 66, 72 and 74, and two Auxiliary Air Force squadrons, Nos. 602 and 611.

The Soviet Union, distrustful of the Imperialist powers and unaware that Hitler's main target was the Soviet Union, signed a non-aggression pact with Germany on 21 August 1939. War, now, was inevitable. On 23 August 1939, the regular, auxiliary and reserve forces of Great Britain were mobilised for war.

A bent No. 72 Sqn Mark I at RAF Acklington. Pilots converting from biplanes had many minor taxying mishaps because of the Spitfire's narrow track undercarriage, long nose and high torque.

No. 19 Sqn, the first Spitfire squadron, 'scrambles' for the Press on 4 May 1939. In 1940, the parachute packs would be in the cockpits.

The Phoney War

As morning mist soaked up the first light of dawn over Poland on 1 September 1939, the German *Blitzkreig* struck. By the time dawn broke over London and Paris, the British and French Governments had learnt to be fact what they had known to be inevitable. At 11a.m. on Sunday, 3 September 1939, the British and French ultimatum expired.

On 2 September, the 160 Battles of the AASF flew to their war stations in France. The BEF left for familiar grounds in north-east France. By 11 September, all units of the British Air Forces in France were in position. No Spitfire fighters were included. It was policy to deploy only one fighter type, the Hurricane, easier to operate and replace. Spitfires were held in reserve, better suited to home interceptor duties than offensive patrols, and, perhaps, to hide their real performance from the Bf 109 *experten*.

The first eight months of the war was the 'Phoney War' period — the *Sitzkreig*. Nevertheless, there was sporadic air combat and essential work to be done. Although the Hurricanes in France were the focus of the Press, Spitfires first fired their guns in combat on 5 September 1939, but in tragi-comic circumstances: 'The Battle of Barking Creek'. A radar station gave a false alert of an incoming raid. Squadrons scrambled uncontrollably. Spitfires of No. 74 Squadron sighted two 'ME 109s' and shot both Hurricanes down, killing one pilot.

The episode revealed starkly weakness in the control and reporting system before serious bomber raids developed. Procedures were tightened and ground staff trained, Spitfires and Hurricanes flying many hours of boring but essential tactical exercises to hone the ground-to-air co-operation in the following months.

The air defence of Great Britain relied upon the radar chain along the Channel and North Sea coasts. In the months of the Phoney War, Fighter Command refined the alerting system and the tactical placing of squadrons to the highest degree of efficiency. Without the Home Chain radar stations, the Spitfires and Hurricanes could have been squandered on misdirected interceptions of fresh air.

It was the Spitfires of two Auxiliary squadrons that first saw action and claimed the first victories, Nos. 602 (City of Glasgow) and 603 (City of Edinburgh), based at Drem and Turnhouse to

No. 611 Sqn Spitfire Mk I tests its eight machine guns at the stop butts at RAF Digby, in No. 12 Group, in January 1940.

No. 72 Sqn Spitfire Mk I is warmed up in the half-light of a winter day in early 1940 at RAF Church Fenton.

Cartridge links fall from the ejection chutes as a No. 611 Sqn's Mark I tests its guns: there was still little action for the Spitfires.

protect the Royal Naval Base at Rosyth. On the morning of 16 October, several *Luftwaffe* reconnaissance aircraft flew over the Firth of Forth at high altitude, the omen of a bombing raid. The Spitfires were brought to alert.

When the bombers came, nine Junkers Ju 88 fast bombers from *Kampfgeschwader* 30, *Luftflotte* 5, based in Norway, the radar had failed but Red and Yellow Sections of No. 603 Squadron and at least one No. 602 Squadron Section made contact. At 1430 hrs., Red Section's three Spitfires, led by Sqn Ldr E. E. Stevens, took off on patrol from Turnhouse. Their view over the Forth deep into the Highlands and south into England was immediate and wide, but their eyes scanned the black waves of the North Sea. In the autumn haze, they spotted a twin-engined aircraft to starboard, and throttled hard towards it. Stevens turned head on to the bogey, assessing. He climbed his section up in line astern, then dived in, checked the black crosses, and pulled under the Ju 88's belly and squeezed his gun button, raking the bomber's fuselage. His Numbers 2 and 3 followed in perfect execution of the pre-war training attacks. The bomber's rear gunner stopped firing. The Auxiliaries watched the Ju 88 lose its grip on the air, slowly, and slap into the sea off Port Seton at 1445 hrs. The pilots landed at Turnhouse at 1455 hrs.

Stevens was officially credited with the first Spitfire victory, somewhat obscuring the fact that the first RAF fighter interception of the war had been an unco-ordinated affair and the Ju 88s had succeeded in bombing and damaging cruisers in the Forth.

On 28 October, patrolling Nos. 602 and 603 Squadrons' Spitfires shot down the first German aircraft to come down on British soil since 1918, an He 111 which bellied-in on the Lammermuir Hills. On 20 November, a section of No. 74 Squadron Spitfires, led by Fg Off Measures, intercepted an He 111 at 27,000ft. about 15 miles off Southend. It did not make Germany.

Fighter Command Spitfires slowly accumulated victories over the next few months, but BAFF's Hurricanes got all the glory. For the Spitfire squadrons, the Phoney War period epitomised the 99 per cent boredom factor of war, as they flew routine exercises. It was, however, a vital period of growth, for the Command was not yet fully prepared for war. Squadrons were still re-equipping with Spitfires.

Dunkirk

The real war began abruptly at dawn on 10 May 1940 when Hitler unleashed Germany's forces on France and Belgium. The *Blitzkreig*'s ferocity took the Allies by surprise. On 15 May, the Germans broke through at Sedan. With British fighter units forced to retreat to the west, Fighter Command was called upon to support the BEF.

From 10 May, No. 11 Group's Spitfire squadrons had been flying offensive patrols. No. 66 Squadron scored a victory over The Hague on 12 May and damaged three enemy aircraft over Rotterdam, but the first, long awaited, Spitfire versus Messerschmitt Bf 109 or Bf 110 combats did not take place for two weeks. After the evacuation of the broken BEF from the Dunkirk beaches began on 21 May, Fighter Command increasingly encountered the *Luftwaffe* in force. Home-based Spitfires and Hurricanes were ordered to concentrate upon preventing the bombers getting through t the beaches, at great risk to themselves For Göring, the BEF concentrated a Dunkirk presented the stage to prov that his *Luftwaffe* could smash an arm swiftly. The might of four *Fliegerkor* was targetted on the BEF.

The first Spitfire verses Bf 109 comba took place on 23 May. That morning pre-occupied by the engine problems c their CO, a section of three No. 7 Squadron Spitfires had tangled incor clusively with Bf 109s. Two Spitfire reported back at base, Hornchurch, th Sqn Ldr F. L. White had force-landed a Calais-Merck aerodrome. Three No. 5

Penalty! During the Dunkirk evacuation, Operation DYNAMO, Spitfire pilots began to score in the air. This is RAF Gravesend with No. 72 Sqn Spitfires and No. 601 Sqn Hurricanes. No. 72 Sqn operated from Gravesend in the closing stages of DYNAMO, 1-6 June 1940, then returned to Acklington.

Squadron pilots were detailed to rescue him, one, Leathart, flying a Master. At Calais-Merck, the Master, with White aboard, was just climbing out when Bf 109s streaked in firing. One Bf 109 flew right in front of Flt Lt Al Deere's Spitfire. Deere fired, trying to distract the Bf 109 from the Master. The Bf 109 veered away then shot up in a vertical power-climb. Deere fired again, smashing the engine. The *Emile* crashed into the tideline.

Above, his wingman, Allen, was heavily involved. Deere climbed hard. Two Bf 109s crossed his line of flight, turning. Deere followed. A long burst ripped pieces off one Bf 109 which rolled and dived to the earth, exploding. The unprotected Germany leader dived away. The German was obviously combat hardened. He turned into Deere's attack. Deere riposted, but his guns clacked silently. He had to break off the engagement. Just above the tree-tops, he climbed hard and escaped into low broken cloud.

On 23 May, No. 92 Squadron fought and destroyed six Bf 109s between Calais and Dunkirk, and lost only one Spitfire. Later that day, they tangled with a large formation of Bf 110 twin-engined *Zerstörer*, knocking down 17 for the loss of three! The game was going in the Spitfires' favour. Again, later that day, No. 54 Squadron, patrolling at full squadron strength for Dunkirk-bound bombers, attacked a group of such 'trade' but were bounced by Bf 109s. Countering swiftly to make up for their mistake, they took out no less than nine Bf 109s for no loss — but the bombers pressed on to Dunkirk. Such combats greatly improved morale. The Spitfire rapidly gathered a formidable reputation.

Some squadrons were in action much more than others, and some pilots flew until exhausted. The air fighting was especially tough as the RAF fighters ignored the enemy fighters to strike the bombers, at high risk to themselves. Nearly all Fighter Command squadrons were involved at some point, squadrons being rotated. Air battles ranged from sea level to over 30,000ft. Bombers were intercepted away from the beaches. The troops were embittered by this apparent non-presence of the RAF by day and night. They were very wrong.

The evacuation brought to England 338,000 British and Allied troops. The RAF claimed 262 enemy aircraft destroyed during the evacuation operation; German figures show 132.

The already latent 'Spitfire myth' grew — the Hurricane units had been decimated in France and, although the comparison was unscientific, the Spitfires were victorious over the Channel. Nevertheless, the loss over Dunkirk of 229 fighters comprised 67 Spitfires and 162 Hurricanes. The scores were not seen with jubilation at Fighter Command HQ. To lose so many in just 14 days — 16 aircraft or over *one squadron* on average a day — reflects the intensity of the air fighting and the drain upon Fighter Command. The loss of 80 pilots was more serious.

Battle over Britain

On 17 June 1940, France fell. Britain was alone. The RAF was given a brief respite. The Dunkirk fighting had been very tactically limited, offering few lessons. The RAF had a hard task.

After the fall of France, it was not until 16 July that Hitler ordered preparations for the invasion of the United Kingdom to be made — Operation SEALION — ready for mid-August. Neither the German army, air force nor navy was prepared or eager. Invasion vessels and barges were collected piecemeal and plans for the landings and thrusts inland from the Channel coast were rapidly prepared. Both the German army and navy feared the consequences of the British Home Fleet opening up on the invasion fleet, while they were in no doubt that their hastily prepared invasion, covered at extended range by the *Luftwaffe*, would be placed in severe jeopardy unless the RAF were first vanquished. To this end, the *Luftwaffe* mounted the campaign that has become known as the Battle of Britain, and, while the tactics changed as the *Luftwaffe* failed, this was always the aim.

The Germans had quickly made up their losses during the Battle of France and could deploy 3,000 aircraft for the assault on Britain. The RAF had just 800 fighters. These were deployed in four groups. There were just 19 Spitfire and 29 Hurricane squadrons.

The first phase of the Battle began on 2 July with small scale raids and continued, until 12 August. It was essentially a phase of skirmishing. The attacks on shipping and ports rose in frequency and intensity throughout July and into August. The first real clash was a *Luftwaffe* victory. The target was an east-bound convoy in the Channel on the evening of 7 July. Bf 109s flew a sweep ahead of the bombers and encountered No. 54 Squadron, destroying two Spitfires and damaging another. As dusk fell, Nos. 64 and 65 Squadrons rose to intercept the Do 17 attack on the convoy but met another Bf 109 sweep, losing three Spitfires and their pilots. Although two Bf 109s were claimed, the *Luftwaffe* lost no aircraft.

On 10 July air activity increased as the good flying weather returned, and Fighter Command flew 609 sorties that day, losing six fighters, but claiming four Bf 109s, six Bf 110s and five bombers. No. 74 (Spitfire) Squadron and three Hurricane squadrons intercepted 26 Do 17Zs preparing to attack a convoy in the Thames estuary. The escort was Bf 109s and 110s. One Do 17 was destroyed in collision with a Hurricane which was also lost; one was shot down by Hurricanes, and four were damaged by Spitfires, while the Hurricanes claimed four fighters and the Spitfires two with seven damaged. No other RAF fighter fell.

This pattern of operations continued, with some violent clashes between opposing fighters until the eve of *Adler Tag* (Eagle Day), 10 August, the day se for the invasion. The *Luftwaffe* had cor sistently suffered losses disproportionat to the results and only on two days wer RAF losses heavier. During this phase the Spitfires and Hurricanes average 400 sorties a day, honing the radar detec tion and fighter control network an gaining invaluable experience ready fo the main onslaught while the *Luftwaf* squandered its efforts.

Bad weather on *Adler Tag* forced i postponement for three days to 1 August. Preparatory to the landings, th *Luftwaffe* was to launch swift, decisiv attacks with the objectives of destroyin the coastal radar stations, neutralisin RAF fighter bases, disrupting RA ground organisation and bringing th Spitfires and Hurricanes to final battl The effort was to be sustained at max mum prejudice for three days, wit attacks extended to London's airfield on day two and sweeping the remain ing RAF fighters from the sky on da three. In this *Blitzkreig*, air superiorit would be rapidly gained. It was war to timetable.

As a prelude, on 12 August, Bf 109 an 110 fighter-bombers and Ju 88s took ou Ventnor radar station and damaged thre

A No. 66 Sqn Spitfire Mk I approaches to land at RAF Gravesend from over the Thames in September/October 1940 with a No. 501 Sqn Hurricane and the Spitfire of No. 66 Sqn's CO, Sqn Ldr Rupert Leigh.

No. 66 Sqn/No. 421 Flight Mark I at RAF Gravesend in September 1940. No. 421 Flight was formed within No. 66 Sqn at the express wish of Churchill to fly high level armed reconnaissance over the French coast. It was renumbered No. 91 Sqn in 1941.

Groundstaff re-arm a No. 66 Sqn Spitfire Mk I at RAF Gravesend, September/October 1940.

others, Do 17s attacked Lympne while Ju 87s attacked a Thames convoy in the morning. In the afternoon, Bf 109s and 110s bombed Manston — just as No. 65 Squadron's Spitfires were taking off — and put it out of action. Later, Hawkinge and Lympne were badly damaged. But losses were indicative: almost 4 per cent.

On 13 August, the *Luftwaffe*'s attacks developed late, postponed by poor weather, but 74 Do 17s had already set off, unescorted. No. 74 Squadron's Spitfires hit them as they began to bomb Eastchurch airfield. Nos. 111 and 151 (Hurricane) Squadrons then set upon them. A vicious fight developed over the Thames estuary, with five Do 17s falling. In the afternoon, the Spitfires of No. 609 Squadron caught 52 Ju 87s making for Middle Wallop and shot down six, while Hurricanes took on the Bf 109 escort. Within minutes, 120 Ju 88s began a run on Southampton and Middle Wallop. Causing only light damage, they were viciously assaulted by the Spitfires and Hurricanes. Seven Ju 88s and four Bf 109s fell. By the end of the day, No. 609 Squadron's Spitfires had claimed 13 victories. In total, the *Luftwaffe* lost 43 aircraft, its highest loss to date.

The English weather clamped down on 14 August, and the *Luftwaffe* could launch just one major raid, on Manston. The campaign was not running to schedule but the *Luftwaffe* made the first of a series of serious tactical errors. Believing that the RAF had been forced to bleed its northern reserves, the *Luftwaffe* High Command ordered heavy attacks on airfields and radar stations in the south and directed *Luftflotte* 5 in Norway to launch attacks on Scotland and Northern England to destroy the remaining reserves. 'Black Thursday' cost the Germans 72 aircraft and *Adler Tag*'s further postponement.

Thursday, 15 August 1940 was a turning point in the Battle of Britain. The *Luftwaffe* flew 1,786 sorties attacking targets in the south-east of England throughout the day, but the two heavy

raids it launched on the north-east were the last it launched in strength from Scandinavia.

At 1208 hrs., No. 13 Group's operations room plotted a raid heading south-west towards the Tees and Tyneside areas from the North Sea. The bombers were 65 He 111s of *KG* 26 escorted by 35 Bf 110s of *ZG* 76, from *Luftflotte* 5 at Stavanger. Seven No. 13 Group squadrons were soon heavily engaged. But, 90 miles south, the rest of *Luftflotte* 5's bomber force, 50 Ju 88s of *KG* 30 from Aalborg, Denmark, were heading for RAF Driffield. The defence against this raid fell to No. 12 Group.

No. 616 Squadron's Spitfires were scrambled from Kirton in Lindsey at 1300 hrs. and ordered to Hornsea. The pilots were lunching and the flights were changing over readiness duty, and both flights were able to scramble. At 1305 hrs., No. 624 Squadron's Defiants from Kirton were ordered to patrol a convoy in the Humber. At 1306 hrs., No. 73 (Hurricane) Squadron was ordered to assign B Flight to Church Fenton's defence and A Flight to a second convoy. At 1310 hrs., No. 13 Group contributed the Blenheims of No. 219 Squadron from Catterick. The defence had mustered.

The enemy entered the area at 1315 hrs. No. 616 Squadron was ordered north. Ten miles off Flamborough Head, it encountered the enemy, who was flying in irregular formation at 15,000ft., unescorted. The squadron turned east to bear, and attacked. The Ju 88s immediately broke for cloud, with little return fire. No. 73 Squadron's A Flight of six Hurricanes joined the fray. The raiders included Ju 88C heavy fighters in lieu of escort fighters, but they were ineffectual. The RAF took a heavy toll. Just 30 bombers got through to bomb Driffield, Bridlington and Scarborough but, by 1420 hrs. the last enemy bomber had gone from the north-east. No. 616 Squadron alone claimed eight bombers destroyed.

The *Luftwaffe*, however, returned in force in the south on 16 August to attack radar stations and airfields, but lost 46 aircraft compared to RAF losses of 23 fighters. The heaviest German loss was of nine Ju 87s shot down by Hurricanes, and the heaviest RAF loss of five No. 266 Squadron Spitfires to Bf 109s. On 17 August, the Germans licked their wounds. The next day brought home to the *Luftwaffe* that the Spitfire and Hurricane were far superior to any fighters they had so far encountered in Europe and that the standards of professionalism and courage displayed by the pilots were equal to their own.

On 18 August, the *Luftwaffe* launched major attacks upon the sector airfields, to which they mistakenly believed the RAF's fighters were tied. They had badly misunderstood the Fighter Command system of direction. Biggin Hill and Kenley were hammered in the morning, but No. 43 (Hurricane) Squadron diverted a raid on Croydon. In the afternoon, Ju 87s attacked Ford and Thorney Island airfields and Poling radar station. They were decimated by No. 43 Squadron's Hurricanes and the Spitfires of Nos. 152, 601 and 602 Squadrons — 34 fell, 14 of them to the guns of No. 602 Squadron. Further east, Spitfire and Hurricane squadrons had fallen on a pack of Bf 110s escorting bombers and hacked down no less than 14 and damaged five. In other combats, seven He 111s and 14 Bf 109s were killed. The Germans lost 65 aircraft on 18 August. During the Battle, only 15 August had brought higher losses. However, on 18 August, the RAF lost 43 aircraft, including seven on the ground, its highest loss on one day in the Battle.

The *Luftwaffe* had failed in their tasks, in the face of the co-ordinated, hard-pressed Fighter Command attacks, to reduce the RAF in the allotted period. The *Luftwaffe* reassessed the situation. The Ju 87 was withdrawn and the role of the Bf 110 minimised. Göring, blaming the Bf 109 units for the losses, believing that they were failing in their escort task, ordered them to fly close escort to the bombers, thereby denying them initiative. The *Luftwaffe* also changed its tactics, beginning a series of night raids, with limited activity by day.

In the partial respite between 19 and 24 August, occasioned by the *Luftwaffe*'s heavy losses and new tactics and promoted by the cloudy weather which closed in, Fighter Command took the opportunity to replace tired squadrons in No. 11 Group with fresher units from the north. No. 616 Squadron — one of six Auxiliary Air Force Spitfire squadrons to fight in the Battle — was one.

No. 616 Squadron was posted to Kenley in No. 11 Group on 19 August and thrown into the Battle proper. At Kenley, it exchanged its Spitfires with No. 64 Squadron who then replaced No. 616 Squadron in the Church Fenton Sector. Such rotations and exchanges were part of the overall Fighter Command tactics. No. 616 Squadron had 22 pilots and 18 Spitfires. Just three weeks and eight new pilots and 12 replacement Spitfires later the squadron had only three pilots fully fit to fly, but in those three weeks, with only three days rest, it destroyed 2[illegible] enemy aircraft. The squadron faced the raids on the airfields.

On the evening of 22 August 1940, the squadron was at readiness. At 1845 hrs. it was scrambled, managing to get 1[illegible] aircraft into the air, an usually high number. At about 1930 hrs., the squadron arrived over Dover at 12,000ft., just as the second large fighter free-chase of the evening was leaving the Kent coast. The squadron was immediately bounced by about a dozen Bf 109s. P/O. Hugh Dundas, one of Green Sections' pilots, was hit many times. The Bf 109 pu[illegible] shells and bullets into his engine and glycol tank, smashed his controls and jammed the hood. He spun down out of control from 12,000ft. to below 400ft before he managed to free the hood and bale out from the burning aircraft. His 'chute jolted open four seconds before he hit the ground.

The Battle's third phase began on 24 August with widespread and heavy attacks on fighter stations inland. This was the 'critical' phase for the RAF with No. 11 Group severely depleted and exhausted within days, and squadrons from other Groups being drafted in

Three No. 66 Sqn/No. 421 Flight Spitfire Mk Is take-off from RAF Gravesend, September/October 1940. The RAF still flew Vics. The pilots realised that their tactics needed revision, but not at the height of a major battle.

Sqn Ldr Darley, No. 609 Sqn's CO in his Spitfire, displaying four victory symbols. A toughened glass windscreen — costing 6 mph — was introduced in early 1940 to protect pilots along with 73 lb of steel armour plate.

to replace squadrons which became non-effective. On 25 August, No. 616 Squadron alone lost seven Spitfires in combat against Bf 109s over Essex, with two pilots killed and four wounded. Only Denys Gillam, a flight commander, scored. Their base, Kenley, was bombed to a shambles and on 3 September the severely depleted No. 616 Squadron was withdrawn to Fowlmere in No. 12 Group.

The free-chasing fighter battle climaxed between 30 August and 6 September with the RAF losing the equivalent of one squadron a day. The RAF lost 103 pilots killed and 128 seriously wounded in the two weeks after 24 August, a quarter of the trained establishment. Replacement pilots were arriving at squadrons with no more than 20 hours in Spitfires — barely enough to know how to fly it let alone fight it, and some had never fired the guns. Sadly, several 'rabbits' were lost on their first sortie, bewildered. While the RAF's potential decreased alarmingly, the *Luftwaffe* increased the number of fighters it sent. The coastal radar chain was functioning well below a safe level. Fighter Command was not yet defeated, but defeat was in sight.

Fighter Command was saved by the *Luftwaffe*'s final tactical blunder: an all-out bombing assault on London itself to force Fighter Command to squander its last reserves in defence of the capital.

Fighter Command was allowed scant time to reform but non-effective squadrons, such as No. 616 Squadron, posted to Kirton in Lindsey on 8 September, were posted north to make room for fresh units. Trained reserves were posted in: No. 616 Squadron's effective personnel were transferred to other front-line squadrons to make good losses. After a 70-strong attack on London docks on 5 September, the campaign began in earnest on 7 September. From 1500 hrs. until dawn next day, 625 bombers unleashed their loads on the docklands. As Kesselring foresaw, the squadrons north of London retaliated but it was not the spent force he had anticipated. Although No. 11 Group had been decimated, No. 12 Group, held in reserve, was still very capable. Based further north, they had time to form squadrons up into 'Big Wings' which could hit German formations in force. On 9 September, only 90 from a force of 200 bombers reached Greater London and 28 were shot down, most by Spitfires. Day and night raids now became the pattern, but Spitfires could do little by night.

On Sunday, 15 September, the *Luftwaffe* attacked in two waves with a total of 200 bombers. 'Black Thursday' was now succeeded by 'Black Sunday'. On this day they achieved what they had attempted to do all along! Fighter Command had to commit all its squadrons to countering the first wave. There were no reserves left. But the *Luftwaffe* missed the opportunity. It was fully two hours before the second wave came. The Spitfires and Hurricanes were re-armed, refuelled and waiting. Of the bombers, 148 bombed London but 56 were shot down. It was a stunning defeat and the *Luftwaffe* could no longer stand such attrition.

Heavy cloud obscured the UK in late September and attacks diminished. A state of stalemate turned in favour of the RAF as it rebuilt its forces while the *Luftwaffe* dared not fly in force over Britain.

However, Spitfire production was attacked at source with the Supermarine works being bombed on 24 and 26 September and then the new Westland works at Yeovil on the 30th. It was too late. The *Luftwaffe* turned to night bombing entirely from early November. The Battle of Britain was over — a narrow victory for Fighter Command.

Through the Blitz of winter 1940-41, *Luftwaffe* night raids devastated cities almost at will. Such was the inadequacy of the RAF's main night fighter, the Blenheim Mk IF, that most of the few early successes fell to Spitfires and Hurricanes, mainly by chance on clear moonlit nights. Although designed as a day and night fighter, the Spitfire did not shine at night. Spitfire squadrons that operated at night could do little. No. 616 Squadron from Tangmere several times patrolled Southampton and Portsmouth, six or eight Spitfires stepped up above 12,000ft. at 1,000ft. intervals to avoid AA fire. Their downwards vision restricted, the pilots saw only bomb fires, AA flashes, exhaust flare, clouds and bombers fleetingly silhouetted. Winter weather exacerbated problems, and pilots isolating bombers too quickly became disorientated, losing the bomber among clouds while they concentrated on their instruments.

Airborne Interception (AI) radar-equipped Beaufighters, operational from September 1940, linked with an effective ground control interception (GCI) system from January 1941, slowly transformed the RAF's night fighting capability. The raid on London on 19/20 April 1941, in which 24 out of 700 bombers fell to fighters, was the last major raid. The RAF could now turn to the offensive.

PR

Adaptable, stable, fast and with good altitude, the Spitfire was an ideal photo-reconnaissance platform, and, in several marks, served in this capacity in almost all theatres of World War Two, often preceding the arrival of fighter variants, as in Malta, Egypt and Burma. Spitfires at first began PR operations from Heston and *Armée de l'Air* airfields in spring 1940, carrying out important surveys of Northern Italy, Belgium — without the Belgian Government's knowledge — and as far as Stettin, beyond Berlin on the Polish border! Throughout the Battle of Britain, PR Spitfires continued their invaluable work. For the pilots, it was an exacting but rewarding task, as Group Captain H. C. Daish, OBE, RAF (Retd.) recalls:

My first real contact with a Spitfire was in February 1940 when I was posted to the Photographic Development Unit (PDU) at Heston, which had been a pre-war civil airfield and was now an RAF Station. I had, of course, seen these beautiful aircraft flying but had no expectation of ever having a chance to fly one. After all, I was a Fairey Battle pilot trained in low-level bombing, not a fighter pilot. However, one of the requirements for the pilot of a Spitfire doing photographic work was pilot navigation and this was something all Battle pilots had, whereas fighter pilots, who would normally have been the logical choice for this work, mainly relied on radio to get them to their target and home again.

On 26 February 1940, I made my first flight — and what a joy it was. Taking off required one to hold the control column hard over to counteract the tremendous torque, but as the speed built up this was gradually neutralised until the aircraft took to the air and became as light as a feather to handle. She was smooth and gentle and one could 'feel' her much better than heavier aircraft. She responded immediately and positively to the slightest and lightest touch on the control column, climbed better than any aircraft I had up to that time flown and was very much faster. Perfection in the air seemed to me to have arrived and I gave thanks to her designer for the perfect flying machine!

The Spitfires flown by the PDU (which was renamed the Photographic Reconnaissance Unit after the fall of France on 17 June 1940) were camouflaged in Duck Egg Green. All protuberances — rivets, screw heads, etc. — were countersunk and the fuselage highly polished, which added some extra mph to the speed — a very necessary requirement as our

The narrow cockpit and long nose restricted downward vision, making placing the Spitfire over the photo-target difficult. Sidney Cotton, commanding the PDU, designed the bubble — posed by Daish — to enable the pilot to see below more easily.

Spitfires were unarmed and we could, therefore, not afford to mix it with the Bf 109s In any case, as Recce pilots our job was to get the results of the Recce home and this extra speed enabled us both to outclimb and outrun enemy fighters, the latter being the more important. In lieu of guns we had cameras fitted in the wings which we could operate from the cockpit.

Our work in those early days necessitated flying at 30,000ft. and certain sorties had to be abandoned if we developed a condensation

PR Mark IA N3071, flown by Flt Lt 'Shorty' Longbottom, at Seclin, near Lille, the first Spitfire PR sortie. Unarmed, with an F.24 camera in each wing, it is an overall wan green — Camotint.

X4492 was converted from a Mark IA to a PR Mark IV (Merlin 45) with Types C then F camera configurations. Flown by No. 13 Photo Survey Sqn, RCAF, from Rockcliffe, Ontario, it is seen on 11 August 1943.

trail which would expose what we were doing. Later, of course, we operated at heights very much in excess of 30,000ft. but these extra heights did not seem to bother the Spitfire unduly and she always seemed to answer whatever calls we made on her. At that early stage of the war we operated from various airfields in France as well as Heston but, again for security reasons, we kept the aircraft in the hangers and wheeled them out just before take-off. This was done to avoid the risk of them being seen or photographed by the enemy and so giving away our very existence.

After the fall of France, one of the main tasks of the Unit was to keep an eye on the European ports from Norway to Spain. At first this was done from height but it became necessary, particularly in bad weather when clouds prevented high-level photography, to carry out some low (very low!) level photographic reconnaissances of the ports, and the fast Spitfire could not have been bettered for this task. She was fast and steady near the ground/sea and made a good platform from which to photograph shipping or barges or any other signs of possible invasion preparations. For this purpose we had long focal length oblique cameras mounted in the fuselage.

Poor old Spitfire, like most aircraft in the RAF, sooner or later experiments and war requirements necessitated ever more and more loading and it was not long before extra fuel tanks were added to ours to give enough range to fly deep into Germany beyond Berlin. They were heavily overloaded on take-off but once airborne were still a joy to fly.

It was a sad day when I was posted as I never again had a chance to fly a Spitfire, my last flight being on 5 December 1940. Nevertheless, my memories of the great pleasure I had in flying them still gives me a thrill whenever I see one, in the air or on television or in a museum — God's gift to pilots!

The Spitfire PR Mk X (two-stage supercharger Merlin 77), developed via the PR Mark XI from the Mark IX fighter, had a pressurised cockpit.

A few low-level PR Mark XIIIs were converted from Mark V fighters, with low-altitude-rated Merlin 32s. A detachment, No. 542 (PR) Sqn operated from North Russia, in early 1944, photographing *Tirpitz* in Alten fjord from low altitudes to gather intelligence for the bombing attack that disabled her.

Finger Fours

In 1941, Fighter Command turned to the offensive over France under the aggressive new leadership of Air Vice-Marshal Trafford Leigh Mallory. The objectives were to maintain the momentum of the war in North-West Europe; to drive the *Luftwaffe* on to the defensive and force the Germans to commit disproportionately large numbers of units to France rather than deploy them to the Middle East and Africa; and to flush the enemy into the air and so kill him in quantity. There was as yet no strategic master plan.

Squadrons were small tactical units ideal for flexible interception duties, but wings were ideally suited to offensive operations. The components already existed at sector level, the sector squadrons. In essence, sectors were given a wing commander (flying) charged with leading the squadrons offensively in battle, rather than treating them tactically individually and corporate for administration only. Spitfire wings formed at Biggin Hill, Hornchurch and Tangmere, Hurricane wings at Kenley, Northolt and North Weald, and mixed wings at Duxford, Middle Wallop and Wittering. The wing leaders — Douglas Bader, 'Sailor' Malan, 'Al' Deere, famous names already — began to formulate the tactics that became standard for the next three years. This was part of the wings' function, a small but vital link to the offensive campaigns that opened in 1943. The major problem facing Fighter Command was actually getting the *Luftwaffe* to fight.

The Spitfire and the Hurricane remained the mainstay of Fighter Command. By 1 April 1941, all Spitfires had been re-equipped with the Mark II which had the edge over the Bf 109E. Introduced in the closing stages of the Battle of Britain in August 1940, first with No. 611 Squadron at Digby, it was basically a Mark I built by Castle Bromwich, and lacked the finish of the Mark I but had a 1,175 hp Merlin XII instead of the Mark I's 1,030 hp Merlin II or III, marginally increasing performance.

In this offensive mood, the wings began rewriting fighter tactics — or rather, rediscovering old lessons. The 'Vics' and 'line astern attacks' of the pre-war Fighting Area Tactics were suitable for attacking single or formated bombers but were lethal to the attackers against fighters.

Individual squadrons were vulnerable, therefore one squadron flew low either as escort to bombers or as 'sweepers', one flew at medium altitude to protect these aircraft and one flew high as top cover against attack from high altitude, thus covering a large volume of sky. With several wings, they were stacked. Such formations allowed wing leaders to deploy the maximum force flexibly.

Several operational commanders introduced an equivalent of the German *rotte* to Fighter Command; it had, in fact, been flown in 1917-18. It was named the 'finger-four' and consisted of four aircraft flying in the same relationship to each other as are the fingers of an outstretched hand. The aircraft were staggered in height, to give greater cover and flexibility. It became the standard

No. 72 Sqn Mark IIB refuelling at Gravesend, spring 1941. It has the B wing — two 20mm. cannon and four 0.303in. machine guns. Cannon-armed Mark IBs, equipping No. 19 Sqn in 1940, had suffered frequent cannon stoppages because the installation was then imperfect.

Fighter Command formation, although all squadrons did not adopt it until 1943.

The 'finger-four' was based upon two two-aircraft sections. The leader of each section had a wingman who guarded his tail and supported him if they became isolated. It gave total cover of the sky, and was both superbly defensive and offensive. Three such units formed a squadron, staggered at height intervals and spread wide, a very different kind of tactical unit to the closely formated squadrons of 1940. Wg Cmdr Douglas Bader's Tangmere Wing — Nos. 145, 610 and 616 Squadrons — probably first flew the 'finger-four' in combat, on 8 May 1941. The formation had to retain cohesive defensive and offensive integrity at all times. Fighter combats took place in over 500 cubic square miles of sky at over 250 mph, and it was easy for individuals to become isolated, suddenly sundered from the nucleus of their squadrons and 'finger-fours' as the fighters whirled out like the Big Bang.

A wing's duties were bomber escort operations, offensive fighter sweeps and hit-and-run raids. The problem of getting the *Luftwaffe* into the air took greater effort. Various types of operations were devised. In *Circus* operations, a small number of medium bombers as the bait were escorted by several squadrons of fighters as the killers. Results were random. *Rodeos* and the smaller *Ramrods* and *Roadsteads* were fighter sweeps by numbers of wings or squadrons, attempting to flush the *Luftwaffe* into the air. Airfields were attacked, transport shot up. Air superiority was gained. But still the *Luftwaffe* refused major battle.

Three aces from No. 452 Sqn, the first Australian Spitfire fighter unit formed in the UK, becoming operational on 22 May 1941: From left — flight commanders 'Bluey' Truscott and Paddy Finucane and CO 'Throttle' Thorold-Smith at Kenley in 1941. Truscott and Finucane chalked-up a great number of victories in a short space. All three were killed.

Rhubarbs were limited daylight attacks by small sections of two or four Spitfires or Hurricanes, under cloud or bad weather cover, flown at the discretion of group commanders — usually! — in quiet periods between other missions, to destroy enemy aircraft and attack targets of opportunity on the ground. Over a third were aborted due to unsuitable weather. Very few enemy aircraft were destroyed, but locomotives, rolling stock, small vessels and transportation systems were attacked. The highest loss percentage of any type of air operation in 1942-43 — approaching 10 per cent, most to flak — is the final word on their success.

Hardly had the Mark II re-equipment programme been completed than the Mark V was introduced. It was a Mark I or II with a Merlin 45-series engine. Its introduction was timely because in May 1941 the improved, stronger, faster Bf 109F appeared. The Mark V became the mainstay of the fighter offensive over North-West Europe until early 1943 and served in large numbers abroad. With better speed, climb and turn, and slightly increased range, it had the tactical advantage of resembling the earlier Marks. The arrival of the cannon-armed Mark VB brought about a major change in Fighter Command's offensive capability.

Throughout the summer of 1941, the Wings and other individual squadrons flew long and hard, but not without important losses. They fought at disadvantage over France for if a pilot were shot down, he was captured, while the *Luftwaffe* could take full advantage of climbing to height and attacking out of the sun. On 9 August, the Tangmere Wing took its new Mark VBs into action, except Bader who considered cannons tempted pilots to open fire too soon, and clung to his Mark VA. The mission was a bomber escort to the Bethune area. Bader flew with Jeff West as his wingman and Dundas and his wingman, Johnson, as his second section — his usual cover. Heavy cloud up to 12,000ft. over France made conditions difficult. A few miles from Bethune, a force of Bf 109s threatened, and Bader led No. 616 Squadron into the attack. Then, No. 145 Squadron, flying top cover, was forced into the fight when more Bf 109s arrived. The fight became confused. Bader shot down two Bf 109s in rapid succession, but by then Dundas, West and Johnson had all become involved in individual fights and were isolated. All four pilots were vulnerable.

All pilots had returned in ones and twos by lunch, except Buck Casson and Bader. A few days later they heard both were prisoners of war, Bader sliced down in a collision in the mêlée. Bader had led the Tangmere Wing for only four months, but his leadership and example were indelible, not only on the Wing but throughout Fighter Command, and his loss was keenly felt.

BIRDS OF PREY

By the end of 1941, the RAF had established air superiority over the Channel and Northern France, but towards the end of the year the balance of fighter air power began to alter. Most UK-based Hurricane fighter squadrons were re-equipped with other types, principally the Spitfire, during winter 1941-42. The Hurricane was almost obsolescent as a day fighter in the Home theatre. The Spitfire now became the RAF's main day fighter at home.

However, on a sweep on 27 September 1941, RAF pilots first encountered a new radial-engined German fighter of noticeably better performance than the Bf

No. 72 Sqn's CO and a flight commander (above) and (above right) the whole squadron show off 1941 fighter pilot fashions: immaculate white pre-war flying suits were rare; thick woollen socks were *de rigeur* in the unheated cockpits; and the 'Mae West' life jacket — named after the well-endowed actress — essential to survive ditching in the Channel.

109F or the Spitfire Mk V: the Focke-Wulf Fw 190A, the fighter the *Luftwaffe* in France had been waiting for throughout the year. Although the pattern of RAF air operations continued into 1942, the formidable Fw 190A wrested from Fighter Command the air superiority it had fought so hard to win in 1941, and many squadrons sustained heavy loss to *Die Würger* — the Butcher Bird — during summer 1942.

To overcome this, the 'stop-gap' Spitfire Mk IX was pressed into RAF service; the RAF could not wait for the 'super Spitfire' then under development, the Mark VIII. A Mark VB or C re-engined with a more powerful, two-stage supercharged Merlin 60-series engine, it was much faster in level flight and the climb than the Mark V, and had a greatly improved ceiling, whilst, again, offering the tactical advantage of being visually similar. In most respects, it was the equal

WAAF flight mechanics strap a pilot into a machine-gun-armed (A Wing) No. 411 (Grizzly Bear) Sqn, RCAF, Mark IIA *Venture I*, RAF Digby, 20 October 1941. Several Commonwealth and Allied Spitfire squadrons formed in 1941.

or better of the Fw 190A, but, although in service in a few squadrons from June 1942, it was not until early 1943 that sufficient had been produced noticeably to challenge the Fw 190, and the Mark V squadrons bore the brunt of the fighting.

Among the squadrons re-equipping with Spitfires in late 1941, were Commonwealth units and the American Eagle squadrons. American volunteers had fought in Hurricanes during the Battle of Britain, but during 1941 three Eagle squadrons formed, Nos. 71, 121 and 133, and were equipped with Spitfires. Over winter 1941-42, the three squadrons flew East Coast convoy patrols, then fighter sweeps, bomber escorts, rhubarbs and other penetrations into France. In May 1942, No. 133 Squadron moved to Biggin Hill and became part of what was regarded as the premier wing.

JUBILEE

The combined operations raid by Allied troops on Dieppe on 19 August 1942 codenamed Operation JUBILEE was intended to prove theories. The RAF's objectives were threefold: close air support for the landing forces, taking out ground targets; achieving and maintaining complete air superiority, keeping the bombers from the troops and vessels and the fighters from the tactical aircraft; and destroying the enemy in the air, the bait being 252 ships close off-shore and hundreds of fighters aloft. Air Vice-Marshal Trafford Leigh Mallory of No. 11 Group controlled the Air Operations. Fighter Command hoped to encounter the *Luftwaffe* in strength and vanquish them.

Leigh Mallory could deploy 67 fighter squadrons. Three were the virtually untested Typhoons of the Duxford wing. No less than 48 were Spitfires. The majority, 42, flew the Mark V — including the first USAAF squadron, the 307th Fighter Squadron, 31st Fighter Group. Operating as top cover, four flew the new Mark IX and Nos. 124 and 616 Squadrons flew the Mark VI. This was a heavy commitment; there were only 59 Mark V squadrons in the UK. In addition, Army Co-Operation Command contributed four Mustang reconnaissance squadrons, No. 2 (Bomber) Group four Blenheim and Boston tactical squadrons and an RAF Fortress unit took part.

Spitfires were needed to form the continuous 'air umbrella' over Dieppe. Each squadron had to fly patrols of at least 30 minutes duration, either in squadron or using strength. Each group of fighters had to overlap. Any gap or 'tear' could be disastrous for the ground forces.

The Allied land forces went in just before 0400 hrs., but were pinned down on the beaches. Just before dawn, 0445 hrs., No. 129 Squadron's Spitfires whistled in low and fired the first shots of the RAF, boosting the invaders' morale. Then came the Blenheims, Bostons and Hurricanes, with Spitfire cover, to take out Dieppe's five coastal batteries. The *Luftwaffe* had been taken completely by surprise, the first aircraft not appearing until nearly 0700 hrs., and only responding in strength by 1000 hrs. But by 0900 hrs., the vital beachhead withdrawal operations were under way. From 1000 hrs., the RAF Blenheims, Bostons and Hurricanes gave intense close support to the ground forces, while Spitfires broke up the bomber formations.

Luftwaffe resistance strengthened at around 1200 hrs. as the general call for assistance was met — the RAF was getting the major battle with the *Luftwaffe*. Leigh Mallory, who had anticipated this, immediately sent four Spitfire squadrons to sweep and intercept the reinforcements. At the same time, 24 Fortresses escorted by four Spitfire squadrons bombed Abbeville airfield, rendering it unusable for hours. During the crucial phases of the evacuation, Spitfires prevented any bombers attacking vessels. By 1300 hrs., the RAF had air superiority.

Continuous fighter cover was required throughout the afternoon. Some Spitfire squadrons flew four sorties at full strength, for the Germans increased their efforts in the evening as the forces withdrew, and Spitfires, Fw 190s and Bf 109s continued to fight over the Channel until 2100 hrs.

More sorties were flown that day by the RAF than on any so far in the war, 2,339, of which Spitfires flew the vast majority. The Spitfires had won and maintained complete air superiority over Dieppe and had fought the bulk of the

Spitfire (Experimental) N3297 at Boscombe Down in October 1941 after fitting with a two-stage supercharger Merlin 61, and four-blade Rotol airscrew, and a larger oil cooler and supercharger intercooler in a new rectangular fairing below the port wing. N3297 had been the sole Mark III (Merlin XX), the first major design improvement. Considered by Beaverbrook's Ministry for production in the USA, it was superseded by the Mark V (Merlin 45), which could be produced with existing jigs.

No. 121 (Eagle) Squadron, flies over the Stars and Stripes.

A Mark VB AV:R BM590 of No. 121 Sqn, the second Eagle squadron, manned by American volunteers.

Luftwaffe in France and the Low Countries. It went some way to mitigating the disaster on the beaches, but the cost had been dear: for a total of 48 *Luftwaffe* aircraft destroyed, the RAF lost 106 aircraft, 88 of them Spitfires.

WEMENDUM

The work of the Maintenance Units (MU), Servicing Units (SU) and Civilian Repair Units (CRU) was vital as Alex Lumsden recalls:

During World War Two, when Spitfires were being produced at over 400 per month, very many were bent and broken in various degrees. Spitfires with Category B damage — not repairable on site — were sent to CRUs or to one of the RAF's MUs. There, they were stripped for inspection, repaired — sometimes almost rebuilt — modified up to date as appropriate to the Mark, resprayed, given engine runs and then test flown by a Unit test pilot before being returned to service. In those days, before the creation of the Empire Test Pilots School (ETPS) in 1943, MU testing could be a somewhat empirical affair that might have worried operational pilots!

One of the worst problems with early Spitfires was aileron heaviness at high speed. The Supermarine test pilots, having got the ailerons right on particular aircraft, no doubt found it irritating to learn of new troubles at the MUs when the Spitfires had been rebuilt, perhaps with new ailerons or even just one replacement and, quite often, a replacement wing. This could put the cat among the pigeons and I particularly remember that Spitfire Mk VB AD137, the first rebuilt by my MU, required several flights before we achieved the correct balance.

Many MU test pilots were posted on temporary attachment, undergoing a rest after operational tours, as I was, and the atmosphere on the units was relaxed and informal but as professional as could be in the circumstances, typified by the motto of at least two such units:

UBENDUM, WEMENDUM ET TESTUM

Spitfire Mk VB AD137 following rebuild and testing at No. 24 MU Tern Hill: 12 September 1937, with, left to right, Flt Lt Poynor, Flg Off A. S. C. Lumsden, Wg Cmdr James and Sqn Ldr Kerr.

Friends and Allies

Only one of the Eagle squadrons re-equipped with Mark IXs, No. 133 Squadron at Biggin Hill in August 1942, when the Biggin Hill Wing became one of the first to receive the LF Mark IXC. The squadron only flew it for a month. On 26 September, they formed part of the cover for a USAAF 97th Bombardment Group B-17 raid on Morlaix airfield, near Finistère, but the bombers left the formating rendezvous point early. Exeter fighter control vectored the Spitfires after the bombers at full boost. The target area was obscured by cloud and the B-17s, compounding their lack of discipline, got lost. A 200 mph jet stream carried the Eagles and B-17s 100 miles out over the Bay of Biscay. Only two Spitfires, low on fuel, made England, one crash-landing just over the Cornish coast. The other ten decorated the Brittany peninsula. This stroke of ill fortune prematurely ejected the Mark IX from the secret list.

With the arrival of the USAAF in the UK, it was decided to place the Eagles on a more 'regularly defined operational status'. On 29 September, Nos. 71, 121 and 133 Squadrons, RAF, disbanded, re-forming as, respectively, the 334th, 335th and 336th Pursuit Squadrons of the new 4th Pursuit Group, the first two based at Debden, the latter at Great Stamford. Eagles flew their first mission under US Eighth Air Force control on 2 October 1942, their Spitfires bearing US stars. In March 1943, however, having served as the vanguard of the US effort over Europe, they were selected, in view of their experience, to be the first UK-based units to equip with the P-47 Thunderbolt — an exchange of elegant agility for brute force.

Other US squadrons flew the Spitfire in the UK in 1942. The 307th, 308th and 309th Fighter Squadrons, forming the 31st Fighter Group, arrived in England in June 1942 and equipped with Mark VBs, with a few Mark IIAs and VAs for training. The 31st was quickly followed by the 2nd, 4th and 5th FS, 52nd FG, which equipped with Spitfire Mk VBs, the 2nd flying with Fighter Command for a time during the summer. This reverse lend-lease enabled the USAAF to get to grips with the real air war over Europe while waiting for its own fighters to be produced in quantity and shipped to England. However, in October 1942, the 31st and 52nd FG were detailed to take part in Operation *Torch*, the North Africa landings next month. Relinquishing their Spitfires they took ship from the Clyde along with RAF Spitfire pilots on 5/6 November and collected 130 new Spitfire Mk VCs at Gibraltar. Many of their Mark VBs were handed over to the 12th, 107th, 109th and 153rd Observation Squadrons, 67th Observation Group.

By the end of 1942, however, the Fw 190 had proved so superior to the Spitfire Mk V, which still formed Fighter Command's main equipment, that penetrations into France had been curtailed

Development work on the Spitfire was rapid, continuous and comprehensive. Medium altitude: Mark IXC — one of the first delivered, built by Supermarine at Woolston — of No. 402 (Winnipeg Bears) Sqn, RCAF, part of the Kenley Wing, seen on 24 November 1942.

severely. Every experienced fighter leader was needed in this crisis and the Mark IX was pushed into service. The three Royal Canadian Air Force squadrons, Nos. 403, 411 and 421, forming the Kenley Wing, began to re-equip in January 1943, with total establishment reached by October. 'Johnnie' Johnson, overdue for a rest, had been noted as an aggressive, tactically astute squadron commander, leading No. 610 (County of Chester) Squadron during 1942 with great success. He was posted to Kenley as wing leader.

In early April the wing was detailed to cover a Typhoon Wing's withdrawal after a strike in France. Using radar, the fighter controller set the Kenley Wing up superbly above the Typhoons, then vectored it on to an enemy force. Just as the Wing was about to attack, the controller warned of another enemy force behind. Johnson decided to continue the attack. The fight was over quickly and he ordered the Wing home before the second force arrived. For the loss of a Spitfire, the Canadians destroyed six Fw 190s, a highly inspiring introduction for the Wing to the Mark IX.

Throughout 1943, the air war over North-West Europe was changing. The

High altitude: Spitfire HF Mk VII EN474, over Ohio, USA, while being evaluated by the USAAF. The HF Mk VII had a Merlin 71, a pressurised cockpit and long span 'pointed' wings for manoeuvreability in the rarefied air. It was intended for intercepting *Luftwaffe* reconnaissance types — only 16 HF were built, most being F Mark VIIs.

Luftwaffe no longer responded to the *Circuses* and *Rhubarbs*, yet the RAF continued to fly them and continued to lose pilots to flak for little gain. However, other tactics were devised. Typhoons straffed and bombed *Luftwaffe* airfields and other targets, flushing the German fighters into the air — or destroying them on the ground — for the Spitfires that followed.

The major change during 1943 was the beginning of the US Eighth Air Force's day bomber offensive. The Spitfire's short range excluded it from escorting the bombers to their deepest objectives and so from the victories over Germany. It was a period of frustration for the Spitfire squadrons as they watched the focus of the air war shift inexorably to the skies over Germany and out of range, accentuated by the arrival of the USAAF's first long-legged fighter escorts, the P-38s and P-47s. The Spitfires were restricted to Northern France, Belgium and Holland. Although the *Luftwaffe* flew still in wing strength in these areas and the summer brought intense fighting, the Tangmere, Kenley and Biggin Hill Wings scoring outstanding successes, it was relatively unfruitful compared to the skies over the Reich.

However, a new role was being envisaged for the Spitfire during 1943. On 1 April 1943, No. 83 Group formed at

Mark VB W3834 YO-Q *Corps of Imperial Frontiersmen* of No. 401 (Ram) Sqn, RCAF, at RAF Redhill, Surrey, 9 July 1943. The wing tips have been replaced by fairings — 'clipped' — to improve low altitude manoeuvreability.

Redhill, Surrey, for the purpose of giving air support to the planned Allied invasion of France. It was to assist in the tactical air operations essential to make the invasion possible, to provide direct air support during the landings, and then to establish itself on the continent with the invading armies and provide tactical air support throughout the subsequent land battles until Germany was defeated.

Low altitude: Easily identifiable by the new nose is the prototype Spitfire Mk XII, DP845, with a Griffon IIB. The new engine offered more power and greater development potential than the Merlin, originally especially at low-level. DP845 was originally the prototype Mark IV — the first Griffon-Spitfire — which did not enter production, but, modified it served as the prototype for the Mark XII which was rushed into production to counter the low-level, fast Fw 190 'hit-and-run' raids of 1942-43.

RUSSIA

The Red Army Air Force was a major user of the Spitfire, almost 8 per cent of total wartime production being shipped to Russia. When Hitler attacked the Soviet Union on 22 June 1941 in his 'lightning crusade' against Bolshevism, the British and still-neutral American governments quickly realised the vital necessity of preventing the Germans seizing the USSR's immense natural resources. Within a month, the USA had extended Lend-Lease to the USSR, and the UK and USA had jointly agreed to send 200 Tomahawks. The UK also offered Hurricanes.

The Russians had been well-aware of the Spitfire's superlative combat reputation, and, on 20 August 1941, a Soviet test pilot flew a Spitfire at Duxford. Two days later, the Soviet and Naval Air Attachés visited the base. However, when on 1 October 1941, the first of four Protocols was signed covering large quantities of aircraft and war supplies for the USSR from the USA and UK, Spitfires were not included. For Britain, the easiest way to meet these requirements was to ship aircraft brought from America and Hurricanes, obsolescent aircraft by European combat standards. The Soviets were, however, impresssed by the Hurricane and between September 1941 and late 1944, one-fifth of UK and Canadian Hurricane production — 2,952 — went to Russia.

On 1 September 1942, the first Spitfires landed in the USSR, three PR Mk IVs from No. 1 PRU, detached to Vaenga, North Russia, tasked with observing German raiders that threatened convoys. The task over, two were given to the Russians. They had already taken note and, on 4 October

Basrah, April 1943: To speed delivery to the Soviets and ease maintenance/supply problems, 143 Mark VBs from Middle East stocks were brought up to a common modification standard. Recently shipped from the UK, many were well-used. EP495 was built at Castle Bromwich in 1942.

1942, the Soviet Ambassador in London had presented a request to Churchill. The Battle of Stalingrad was raging; the *Luftwaffe* had great strength, but the Russians had few aircraft and could not cover their troops. 'What we particularly need is Spitfires and Airacobras . . .'. Despite pressing demands, on 9 October, Churchill agreed to send 150 Spitfires plus 50 in spares. Mark VBs were quickly allocated from Middle East stocks, and, on 19 October, he sent a rider that they would have two cannon and four machine guns. In early 1943, 143 were handed over to the Soviet Mission at Basrah, and collected by Soviet pilots.

They went into operational service immediately along the Black Sea coast of the Caucasus. Others were allocated to the air defence of Moscow to deal with high-flying Ju 86P reconnaissance aircraft, bringing down one in their first engagement. Later in 1943, pilots of *Jagdgeschwader* 52 were startled to find themselves pitted against 25 Spitfires in a fight near Orel, 200 miles south of Moscow: they were much tougher adversaries than the Yaks and MiGs. The Soviet pilots particularly liked the Hispano 20mm. cannon but were well aware of the age of their mounts.

When Hurricane production ended on 5 September 1944, the only alternative was to send Spitfires, under the Aid to Russia programme terms. Russia wanted the latest version, the Mark IX. From mid-1944, new batches were allocated to Russia from UK Aircraft Storage Units, then, from late 1944, most of Castle Bromwich's production. By April 1945, 1,186 LF and two HF Mark IXE had been delivered via Basrah. The Spitfire was the second most numerous British aircraft in Soviet service — 1,331 Marks V and IX, plus five PR Mark IVs and two PR Mark XIIIs handed over by the RAF in North Russia.

The Spitfire Mk IX was popular with Soviet pilots, more so being new, and was flown by several Fighter Regiments of Air Defence Aviation (IA PVO), including the elite 26th Guards Fighter Regiment, part of the Leningrad Air Defence Force, and strongly influenced the air superiority battle on this and other sectors of the front. Among the few combats recorded, is one in which a Mark IX shot down four Bf 109s in one action in autumn 1944. Some were locally modified, including, as standard with Russian fighters, to two-seat trainers.

Crisis in the Med

MALTA

Until early 1942, Spitfire fighters were reserved for home duties and the offensive over France where the latest *Luftwaffe* types were encountered. However, by January 1942, the British position in the Mediterranean and Middle East had become critical and the first overseas Spitfire fighter deployment was made in March 1942 to Malta, while the first Spitfire squadron became operational in the Middle East in June 1942.

Malta, lying mid-way between Gibraltar and Alexandria, was a crucial supply point for the British in North Africa and the Middle East, while, just 60 miles from Sicily and 200 miles from Libya, its aircraft and submarines could strike Axis supply vessels and aircraft, imperilling Rommel's *Afrika Korps*.

The *Regia Aeronautica* had failed t reduce Malta in 1940. In 1941, the *Luft waffe*, opposed by Hurricanes flown i from Libya and from aircraft carriers also failed. In December 1941, a nev Axis air campaign began. Malta had jus 60 Hurricanes. By January, only Luq airfield remained operational and Mal tese offensive operations had ceasec That month, the Axis lost no supplies

A Mark VB (Tropical) for Malta in a strange environment in January 1942. As the need for the Spitfire in tropical areas became evident, development work rapidly adapted it to the new conditions. Its short range was improved by the slipper tank — essential for flying off carriers Malta-bound — while ingestion of dust and sand, which rapidly wore out the engines, was minimise by the Vokes air filter under the nose and the effects reduced by a larger oil tank and cooler.

Lt R. J. Connor, his crew chief, Sgt W. A. Ponder, and their 309th FS, 31st FG Mark VC at an advanced 12th US TAF base in North Africa, late December 1942. The Group re-equipped with Marks VIII and IX in early 1943.

while the British convoys were cauterised, bombers decimating British supply convoys at will. With supplies, the *Afrika Korps* drove the British, without supplies, out of Cyrenaica. If she were again to be of strategic value, Malta had to have adequate air defence. Spitfires were needed against the *Luftwaffe*'s fighters. There was no time to ship them via the Cape and no possibility of ferrying them across Africa via Takoradi port, West Africa. Although less suitable for carrier take-off than Hurricanes, they too would have to be flown off, at extreme range, by carriers accompanying supply convoys to Malta. It would be a hazardous undertaking.

MALTA SPITFIRE REINFORCEMENT OPERATIONS IN 1942

Operation	Date	Carriers	Launched	Arrived
SPOTTER	7 March	*Eagle* *Argus*	15	15
PICKET I	21 March	*Eagle*	9	9
PICKET II	29 March	*Eagle*	7	7
CALENDAR	20 April	*Wasp*	47	46
BOWERY	9 May	*Wasp* *Eagle*	64	60
L.B.	18 May	*Eagle* *Argus*	17	17
STYLE	3 June	*Eagle*	31	27
SALIENT	9 June	*Eagle*	32	32
PINPOINT	16 July	*Eagle*	32	31
INSECT	21 July	*Eagle*	30	28
BELLOWS	11 August	*Furious*	38	37
BARITONE	17 August	*Furious*	32	29
TRAIN	24 October	*Furious*	31	29

It was not until USS *Wasp* was called in by Churchill that sizeable numbers of Spitfires could be delivered to Malta. Hitherto, the numbers that *Eagle* could deliver were insufficient in the face of high attrition. However, during April the *Luftwaffe* again deployed maximum resources against Malta, reducing her offensive capability to nil and decimating the Spitfires, but on 9 May, *Wasp* and *Eagle* together flew off 64. Next day, in heavy fighting over Valetta Harbour, for the loss of four, Spitfires claimed 23 enemy aircraft. By early June, the air defences of Malta were again numerically adequate but the assault abated as Germany withdrew units to support other fronts.

Throughout July a renewed Axis onslaught was made upon the island and the Spitfires, reinforced yet again, claimed 149 victories for the loss of 36. The Italians and Germans were unable to maintain the momentum into August and Beaufighter offensive operations resumed during the month while in September, with yet more Spitfires, Malta's Spitfires went on the offensive.

In mid-October, the Axis started a week-long offensive to support the supply lines to Rommel but within three weeks the victory at El Alamein and the torch landings brought Malta's crisis to an end. The excellence of the Spitfire as an interceptor, the tight tactical control of Air Vice Marshal Keith Park from July and the indomitable courage of the Royal Navy and Merchant Marine in fighting through supply and support convoys to Malta in the face of overwhelming Axis air superiority, ensured that Malta remained a strategic point in the British war effort.

Two Spitfire Mk VBs and a Hurricane of No. 33 Sqn at Bersis, North Africa.

Mark VC over the Desert. Spitfire fighters turned the Mediterranean air war, but within six months they were pioneering the tactical use of fighters as 'light bombers' — a critical transition in modern air war.

THE DESERT

In North Africa, well supplied, Rommel launched an assault on the Gazala Line in May 1942. The first Western Desert Air Force squadron to fly the Spitfire, No. 145 Squadron, had re-equipped with Mark VBs in May. On 1 June, they flew their first sortie, providing cover for ground-attack Hurricanes. However, the Gazala Line broke in June and the British and Free French Forces were forced to Egypt. The air battles over the desert in June and July led to RAF fighter shortages and the new Bf 109F outmatched the Hurricanes and Kitty-wks. More Spitfire Mk Vs were needed for top cover and offensive work.

No. 92 Squadron had been detailed for North Africa early in 1942, but the transfer of personnel and aircraft by the long route round the Cape occupied several months and had to be followed by a period of training and adaptation to new conditions. No. 92 Squadron was operational in early July 1942. Its first Middle East combat bagged two Ju 87s.

During July and August, Nos. 92 and 145 Squadrons were heavily engaged with Rommel's brilliant armoured push from Bir Hachim to El Alamein. However, in August 1942, General Montgomery took firm command of the Eighth Army. On 31 August, Rommel launched an armoured attack on the British line between El Alamein and the Quttara Depression. Between 1 and 3 September, WDAF fighters, fighter-bombers and medium bombers flew nearly 2,400 sorties. Nos. 92 and 145 Squadrons flew offensive patrols and sweeps over the battle area, straffed enemy supplies and troops and flew many bomber escorts. By 5 September, Rommel's assault had failed, bloodily, crushed on the ground and from the air.

Montgomery prepared his counter-attack. Complete air superiority was vital. The WDAF expended a considerable effort to achieve it. The destruction on the ground of 30 *Luftwaffe* fighters at airfields in the Daba area, ten minutes flying time from the front, contributed considerably, and the Commonwealth air units gained complete fighter superiority over the El Alamein front, in no small measure due to Spitfires, of which there were five squadrons by the time of El Alamein: Nos. 92, 94, 145, 147 and — flown in from Malta — 601.

The El Alamein offensive began on 23 October; the entire WDAF fighter-bomber and medium bomber strength was unleashed, covered by Spitfires. The *Afrika Korps* broke on 4 November 1942. The fighters and fighter-bombers vandalised the enemy as they retreated on the packed coast road to Cyrenaica. No. 92 Squadron gave valuable close support to the Army during the pursuit, and claimed 79 enemy aircraft destroyed. Flt Lt Nevile Duke, a Flight Commander with No. 92 Squadron, laid the foundations for his eventual score of 28 which made him the Mediterranean's top-scoring Allied fighter pilot.

TORCH

Four days after the *Afrika Korps* broke before the Eighth Army, Operation TORCH, the simultaneous landings in Morocco by Anglo-American and in Algiers by mainly British amphibious forces went in on 8 November 1942. Devised to strike Rommel's *Afrika Korps* in the rear and to bring American forces directly into the war with the Axis, it was the second Allied amphibious landing of the European war and the lessons of Dieppe had been well-learnt. Air power was crucial to its success. This time, the landings were supported closely by seven Royal Navy and four US Navy carriers, and priority was given to establishing aircraft on land as rapidly as possible, flown in from Gibraltar. Fighter resistance was anticipated from the Free French in Algeria, who had 55 Dewoitine D.520s, capable fighters in the right hands; FAA units were detailed to take out their base.

The Spitfire was the predominant type used during TORCH, flown by the RAF, USAAF and, as the Seafire, being blooded for the first time, by the Royal Navy. During the preliminary phases, the carrier fighters — Seafires, Sea Hurricanes and Wildcats — provided cover for the landings. *Furious* embarked 24 Seafires, and Lt G. C. Baldwin, RN, took the first Seafire victory, a D.520 which attacked one of *Furious*' nine Albacore torpedo-bombers.

Within an hour of its capture, No. 43 Squadron flew its 'Hurribombers', for immediate close support, into Maison Blanche, Algiers, shortly followed by No. 225 Squadron's Hurricane fighters. Nos. 81 and 242 Squadrons were the first Spitfire units to land at Maison Blanche, followed during the day by Nos. 72, 93, 11 and 152 Squadron's Spitfires and No. 255 (Beaufighter) Squadron and four RAF Blenheim squadron to form an

Mark VC over Tunisia in Mediterranean theatre Sand/Mid Stone/Azure camouflage.

effective forward air support force. There was, however, little air resistance.

The 31st and 52nd Fighter Groups, USAAF, having disbanded in the UK, collected 130 new Spitfire Mk VCs at Gibraltar and joined the new 12th Tactical Air Force Command for the invasion of Morocco. The 308th and 309th Fighter Squadrons, 31st FG, led by Colonel John R. Hawkes were the first US fighters to land at Tafaraoui, a one hour 40 minute excursion from Gibraltar. As they circled the bombed airfield looking for a long uncrated strip, four aircraft skitted overhead. They looked like Hurricanes and had roundels, and the Americans ignored them, but, as the last Spitfire rolled in, the French D.520s dived in a straffing attack, killing one and wounding several pilots. Scrambling rapidly, Spitfires shot down three of the D.520s. The US Spitfires were again in action next day, when three spotted a French Foreign Legion column with light tank support advancing from Sidi-bel-Abbes. The tanks succumbed quickly to the six Hispano 20mm. cannon and the column retired. Although strong when met on the ground and in the air, French resistance was uncoordinated and lacked real will. Morocco was Allied by 11 November.

The British and American squadrons now based in North-West Africa — the USAAF also disposed one P-40, two P-38 and two P-39 FGs in Morocco — formed the North African Tactical Air Force. It was essential for the aircraft to follow the advance on the ground but there were few suitable airfields. Spitfires, flying as top cover for ground attack and light bomber units or operating as close support aircraft, were limited to ten minutes over the front because of its distance from the airfields. The RAF Spitfires had to crowd on to the furthest forward suitable airfield, Souk el Arba, which the USAAF also used; it was very congested. No. 93 Squadron moved forward to Medjez el Bab in early December, but was straffed efficiently by Bf 109s soon after landing and forced out. During the winter, Souk el Arba was turned into a quagmire by the torrential rain and constant activity.

In January and February 1943, the Eighth Army from the east and the US II Corps from the west pressed upon the *Afrika Korps*, while from the north, Malta again contributed strike aircraft to disrupt Axis supplies from Italy, as well as aircraft, including Spitfires, which joined the NATAF and WDAF in attacking enemy ground forces in Africa. For the first time, Spitfires carried bombs.

In Tunisia, as always, the *Luftwaffe* fought hard. Day bombing raids needed heavy fighter escort. The appearance of the Focke-Wulf Fw 190 in North Africa severely damaged Allied air superiority in Tunisia, enabling the numerically inferior *Luftwaffe* to counter Allied air power. Operating as fast, low-level fighter bombers and close-support aircraft, the Fw 190s flew an incredible number of sorties, attacking Allied supply dumps and ports and supporting the last Axis strongholds to the end. As over France, the Spitfire Mk V was found inferior. The same solution was applied: a detachment of Mark IXs, the Polish Fighting Team under Sqn. Ldr. Stanislaw Skalski, was rushed to North Africa in early 1943 and attached to No. 145 Squadron. In two months, they destroyed more enemy aircraft than any other Polish RAF unit during 1943. In the north-west, No. 72 Squadron was recalled to Gibraltar in February 1943, and partly re-equipped with Spitfire Mk IXs, while the 31st Fighter Group re-equipped with Mark VIIIs and IXs. In the next six months, No. 72 Squadron scored 53 victories, the highest-scoring fighter squadron of the campaign. In February, another Spitfire squadron was formed in North-West Africa, No. 225 Squadron, re-equipping from Hurricanes to Spitfire Mk VCs, although, that same month, No. 682 (PR) Squadron was rendered ineffective when many of its PR Mark IVs were lost in a bombing raid on its base. The NATAF Spitfire squadrons were now operating as wings.

On 7th April 1943, the US II Corps and the Eighth Army linked-up and began the drive north into Tunisia. Allied air power began a total onslaught to annihilate the *Luftwaffe*. Transport bases in Sicily and Italy were bombed and transports attempting the flight were cut down. Rommel's supplies began to dry up. On 18 April 1943, 90 Ju 52s and 50 German and Italian escort fighters were attacked by 12 No. 92 Squadron Spitfires and 47 US Ninth AF P-40s. For the loss of six P-40s and one Spitfire, 77 enemy aircraft were destroyed. It was a crushing victory, 'The Palm Sunday Massacre', during which more enemy aircraft were shot down than on any day during the Battle of Britain. On 22 April, Spitfires and SAAF Kittyhawks shot down 14 Me 323 *Gigantes* and seven fighter escorts. Off Cape Bon that night, the sea blazed with petrol — the *Afrika Korps'* funeral pyre. The final allied offensive had driven the Germans into the Mediterranean by 13 May 1943.

Australia

The rapid Japanese expansion in the Far East reached Australian New Guinea in early 1942, isolating Australia, whose naval and aviation bases and economic resources attracted the Japanese. Before invading Timor island, the last stepping stone to Australia, the Japanese tried to neutralise the important Darwin naval base and port. On 19 February, a heavy carrier air strike was launched on Darwin. USAAF P-40 Warhawks defended the area somewhat inadequately, although the attacks had diminished in frequency and weight by April.

The situation was so serious that the Austrialian Government asked Britain for Spitfires, flown by Fighter Command squadrons and led by Australia's foremost fighter pilot and leader, Sqn Ldr Clive 'Killer' Caldwell. At the end of May, Churchill agreed to send three squadrons and to supply Spitfires sufficient to maintain the force at a strength of 16 aircraft per squadron, allowing for a five per month wastage rate. Spitfire Mk VC (Tropical), as the most suitable for the Australian conditions, were diverted from scarce Middle East allocations. Caldwell finally arrived back in Australia on 26 September 1942, boosting public morale considerably.

Nos. 452 and 457 (Spitfire) Squadrons, RAAF, which had formed in Fighter Command in April and June 1941 respectively, were disbanded in the UK in June 1942. They were very experienced units. No. 54 Squadron, RAF, had conveniently been despatched to the Middle East via the Cape, and was diverted to Australia while waiting for an east-bound convoy in South Africa. The three squadrons were to form No. 1 Fighter Wing, RAAF, sometimes known as the Churchill Wing.

During August, the Wing formed as a nucleus at Richmond, New South Wales, and the first Spitfires were uncrated at Point Cook, in Victoria. The process of forming, equipping, training and deploying the squadrons took five months, an unsurprising delay given the complex logistics involved in the transit from the UK, the size of Australia and the newness of the environment for the pilots and Spitfires. In November, some of No. 452 Squadron's Spitfire Mk VCs, equipment, aircrew and groundcrew, including the CO, Sqn Ldr A. 'Throttle' Thorold-Smith and flight commander 'Bluey' Truscott, both aces, arrived at Melbourne aboard the Sterling Castle. Truscott, however, was given command of a Kittyhawk squadron. The Spitfires were assembled at Point Cook in Victoria from where they were ferried to Bankstown, 12 miles outside Sydney, and thence to Richmond (still a major RAAF base) and Schofields, where the three squadrons were initially based for training. Once assembled, the exacting leadership of 'Killer' Caldwell soon forged individuals into squadrons and squadrons into a fighting wing.

The flight north was an arduous ad-

Jack Newman of No. 451 Sqn on Wg Cmdr Caldwell's Mark VC at Strauss Field, 1943. It was a wing commander's privilege to replace unit codes with his initials.

No. 452 Sqn Mark VC on Strauss Field strip, 1943. In Australia, a Dark Earth/ RAAF Foliage Green/Azure camouflage scheme was applied to many Spitfires.

No. 79 Sqn Mark VC (nearest) and Mark VIIIs, with RAAF serials, operating out of Morotai, after the RAAF went on the offensive from 1944.

venture, across the central desert in midsummer. The Spitfires went first to Batchelor, 120 miles South of Darwin before the squadrons deployed to their operational bases. No. 452 Squadron was based at Strauss Field, 60 miles south of Darwin, No. 457 Squadron at Livingstone, south of Darwin, and No. 54 Squadron at Darwin itself. On 1 January 1943, Caldwell was promoted to wing commander, and the wing became operational. By the end of January, over 100 Spitfires had been delivered to Australia; some were required for training.

Early in February 1943, the Japanese re-opened their bombing campaign against Northern Australia, with minor engagements during that month and into March between Spitfires and the bombers and their escorts. On 6 February, No. 54 Squadron claimed the first Pacific Theatre Spitfire victory when Flt Lt R. W. Foster intercepted a Japanese Army Air Force Mitsubishi Ki-46 *Dinah*, a twin-engined reconnaissance aircraft, 35 miles off Cape Van Diemen. On 2 March, Caldwell claimed his first Australian victories, a *Zero* fighter and a Nakajima B5N *Kate* torpedo-bomber, and Thorold-Smith also opened his scoring with a *Zero*. The aircraft were part of a 16-strong force which attacked Ocomalie airfield, south of Darwin. Sadly, on 15 March, on only the second occasion that he led his squadron into action in Australia, 'Throttle' was killed in action when the Wing intercepted 14 Japanese over Darwin, taking out seven for the loss of four Spitfires.

Sqn Ldr R. E. Thorold-Smith, DFC, CO of No. 452 Sqn (left) and Wg Cmdr Clive Caldwell, wing leader of No. 1 Wing.

There followed a brief respite from raids. The Wing had much to discuss. The early warning radar station on Bathurst island, north of Darwin, was essential to give time for the squadrons to scramble, climb to height and then get up-sun of the Japanese — for the enemy had the advantage of flying with the sun behind them. Caught down-sun in a hard climb by Japanese fighters, the Spitfires would be slaughtered.

Tactics were being thrashed out. The pilots used to engaging Bf 109s, Fw 190s and Macchis in close combat had discovered that the Spitfire was dangerously inferior in maneouvreability to the main Japanese escort fighter — the light, nimble, radial-engined Mitsubishi A6M Zero-Sen, the *Zero* or *Zeke* in Allied code. They were flown by combat-hardened veterans, many of whom had impressive scores. The Spitfire was, however, marginally faster in level flight and much faster in the dive; official permission was sought to wax and polish the Spitfires to gain extra speed, although the harsh climate's effect upon engine power tended to erode the Spitfire's advantage. With advice passed on from the American Volunteer Group — the Flying Tigers — who had fought the Japanese with even more inferior P-40s in China, the Spitfire pilots learnt they had to 'bounce' the *Zeroes*, by diving from height, firing and breaking away in a dive or a spin. Caught in a Spitfire's cannon fire, the fragile unarmoured Japanese fighters disappeared in fragments and fire. It was hardly subtle, but Caldwell gave no points for pilots who died heroically by 'playing fair'. What he wanted was killers. Integrating the early warning system and perfecting the tactics took time, and were dogged by serviceabililty problems and an ignorant Press.

The first major encounter took place on 2 May 1943. A heavily escorted bomber force — 18 bombers and 27 fighters — from the 202nd Air Corps, 23rd Air Flotilla, from Penfui near Koepang, Timor, led by Lt Cdr Suzuki Minoru, headed for Australia over the Arafura Sea. The Bathurst radar detected the force 49 minutes flying time from Darwin. Within five minutes all off-duty pilots had been collected; within 15, the Wing was airborne. Vectored by ground control, the 33 Spitfires climbed at full boost to 26,000ft., with Caldwell at the head of No. 452 Squadron, to a point ten miles north-east of Darwin.

Despite the long warning, things went badly wrong. The raiders, flying with the sun behind them, had a 4,000ft. height advantage. Caldwell did not hesitate: an attack would be suicidal. He withdrew the Wing.

The bombers struck Darwin with precision, but, fortunately, little effect. The inhabitants saw no Spitfires — only an intact formation of Japanese bombers and fighters. Not for the first time in World War Two were the rules of fighter warfare misunderstood by those on the ground. The bombers turned due west from Darwin, instead of returning north, and flew inland along the coast for 30 miles before turning north above Fog Bay at five miles high above the dark blue Timor Sea.

A mile above in the sun's glare, Caldwell held the Wing close to their ceiling, keeping the surprise. Out over the Timor Sea, he quietly ordered No. 54 Squadron to hit the *Zeroes* first, detach them from the bombers and keep them occupied. 54's Spitfires peeled off and dived at

400 mph plus into the *Zeroes*, firing at fleeting targets, and broke off. Seconds later, No. 457 Squadron plunged into the bombers, each Spitfire selecting a target, but the *Zeroes*, superior in number, turned upon No. 457 Squadron. Caldwell, briefly assessing the whirling pattern of the fight below, led No. 452 Squadron into the fray. Over confident in their superior position, the Spitfires engaging the escort got into dogfights down to 7,000ft. while those attacking the main targets, the bombers, became vulnerable. The *Zeroes* did their job well and shot down five Spitfires, two of whose pilots were killed. In addition, in the strain of combat, three Spitfires had engine failure, while on the return flight, five, at extended range, ran out of fuel and force-landed. Five *Zeroes* were confirmed destroyed, but only one bomber.

The Australian Press was merciless in its ignorance: the Spitfire of which so much had been heard and so much expected had so far been singularly unsuccessful. Public confidence plummetted. The pilots' confidence, however, remained high and Caldwell unmoved: it was largely a matter of tactics.

Camouflage net pattern thrown on to a No. 452 Sqn Mark VC at Bankstown in 1943.

On 20 June, the Spitfires intercepted two morning raids. Two Spitfires fell in combat, but nine bombers and five fighters were destroyed and ten were claimed as damaged. On 28 June, four *Zeroes* were knocked down; on another raid, eight bombers and two *Zeroes* were shot down for the loss of three Spitfires — but four more Spitfires were lost through engine failure. By now, Allied bombers in the theatre were attacking Penfui by night, but the USAAF's 380th Bombardment Group, with B-24s, moved into Darwin and became a prime target requiring Spitfire defence.

Flights were detached to advanced landing strips to counter the Japanese aircraft operating increasingly along the Australian coast. Millingimbie, a small island in the Arafura Sea, was one strip used by No. 452 Squadron, three Spitfires at a time being deployed to create the impression of greater strength. Later, during the 1943 Timor campaign, Drysdale in Western Australia was used as a forward base for one or two sections of three No. 452 Squadron Spitfires, guarding against a carrier-strike.

In July, No. 54 Squadron scrambled to intercept a 47-strong raid on Darwin, but only seven of its Spitfires reached the raiders, the rest either grounded or turning back through engine problems. Nevertheless, with tactical acumen, the seven knocked down seven bombers and

Mark VIIIs and aircrew of No. 548 Sqn, RAF, which formed in Australia, along with No. 549 Sqn, RAF, from nucleii from No. 54 Sqn, RAF, in mid-1944 as Darwin's defence units while RAAF Spitfire Mk VIII units fought in New Guinea.

Six Mark VCs of A Flight, No. 452 Sqn, flying from Drysdale strip, Western Australia, on a sortie over Timor in 1943.

B Flight, No. 452 Sqn Mark VCs flying from Batchelor, late 1942.

two *Zeroes* for no loss. If the tactical situation was no longer serious, the supply and maintenance problems were becoming crucial.

Despite the promised arrangements, by late June, No. 1 Fighter Wing was 20 aircraft under establishment, with a higher than predicted loss rate, and the survivors were worn out by the harsh environment. The dust and the long distance needed to taxi from dispersal to the strips, the effects of the intense heat and continual operations on the temperate climate liquid-cooling systems of the Merlins and the heavy use of boost in the climb to altitude and in engaging the *Zeroes* caused rapid engine deterioration. There were several losses through engine failure, both during combat and on interception climbs and returns to base, while the worn-out Merlins were reducing performance, a critical factor in the struggle against the *Zeroes*. There were no replacement engines. Gun stoppages were frequent, worn by use or freezing at high altitudes.

The UK was under strong pressure to supply Spitfires to several fronts, not least the Russian. The situation was critical by the time the first replacements for several weeks were unloaded at Melbourne and flown via Sydney to Darwin. The Wing's old aircraft were handed over to No. 2 OTU at Mildura, New South Wales to train new Spitfire pilots. In total, 247 Spitfire Mk VCs (others were lost in transit), arrived in Australia for No. 1 Fighter Wing in 1942-43 — a measure of the attrition rate.

On 20 August, the Wing secured a notable victory. Early in the morning, three Japanese reconnaissance aircraft appeared over Darwin, the harbingers of a raid. The Spitfires shot all three flaming into the bush. The Japanese sent another. Caldwell himself killed it — his last victory. After that, another was sent, heavily escorted. No. 54 Squadron scrambled to intercept the formation at six miles high, and the *Zeroes* fell on them as they climbed, shooting down three Spitfires, but the squadron evened the scoring by destroying one *Zero* and claiming two so badly damaged as to be impossible to fly back to Timor.

Next month, Caldwell was posted to No. 2 OTU as Officer Commanding and Chief Flying Instructor, to pass on his invaluable experience at source to new pilots. By then, the day raids had slackened. Shortly, they ceased and night raids began, but, after a limited campaign, these too ceased in early 1944. In eight months, No. 1 Fighter Wing had gained effective air superiority over Northern Australia. The arrival of Spitfire Mk VIIIs confirmed it undeniably.

Sicily and Italy

The Western Desert Air Force and the British squadrons of the North African Tactical Air Force joined forces in Malta in June 1943 for the thrust into the 'soft underbelly' of Europe — the invasions of Sicily and Italy. Their first task was to attack enemy communications in June, turning to airfields from early July. Operations escalated, by day and night, with heavy, medium and fighter bombers. Spitfires escorted light bomber squadron sweeps but there was little coherent air opposition despite much fighting. Despite the lack of success of the sweeps over Sicily, the Spitfires rapidly achieved air superiority over and around Sicily.

Typical of the Spitfire wings was No. 324 Wing, No. 210 Group, NATAF, operating out of Hal Far, Malta, and comprising Nos. 43, 72, 93, 111 and 243 Squadrons, flying Spitfire Mk VCs, VIIIs and IXs. In June, 'Cocky' Dundas took command of No. 324 Wing. The Wing flew in the offensive air superiority role, employing the squadrons flexibly. No. 43 Squadron flew its first sweep over Sicily on 15 June, but No. 72 was involved in interception scrambles from Malta to protect the overcrowded airfields and did not make its first sweep until 30 June. Up to three times a day, Dundas led three or four squadrons the 60 miles to Sicily, escorting two or three light bomber squadrons on a sweep. The strain was gruelling and the fighting hard, the Wing claiming 60 victories before the invasion.

During the landings on 10 July, the DAF's Spitfires operated closely in support of the ground forces. Air cover had to be secure. No. 43 Squadron spent D-Day Minus One patrolling the convoys heading for the beaches, without action. On D-Day, elements of No. 324 Wing took off from Malta in the dark to arrive over the beachhead at first light, encountering a few enemy aircraft. On D-Day Plus Two, the Wing landed at Comiso.

After the landings, the Allies held air superiority and there was little enemy air activity by day, though more at night. The DAF's Spitfires flew sweeps and *Rhubarbs*, escorted day bombers on attacks on enemy supply lines north of the battlefront and flew standing patrols over the beachhead and patrols to the bomb-line to intercept German or Italian bombers. It was a forceful demonstration of tactical air power, with squadrons being employed flexibly; No. 43 Squadron, for instance, also flew offensive patrols, while No. 243 Squadron concentrated upon flying heavy bomber escorts. There was still air fighting, too, and, on 12 July, No. 72 Squadron claimed three enemy aircraft destroyed and ten damaged for the loss of two Spitfires and, next day, five destroyed, eight damaged and two probably destroyed, satisfying scores for any squadron. The enemy withdrew to the north-east, and evacu-

This No. 111 Sqn Spitfire Mk IXC, EN364, was shot up during the invasion and, with one undercarriage leg down and the other 10 per cent down, crash landed at Calpequerone, Sicily. From 1943, Spitfire units in the Mediterranean re-equipped with Marks VIII and IX.

Palermo, Sicily, August 1943: Lt John E. Fawcett, USAAF, with his 2nd FS, 52nd FG Spitfire Mk VC bearing the 2nd's 'Regal Beagles' insignia — the 'American Beagle Squadron'.

HF Mark VIII of No. 417 (City of Windsor) Sqn, RCAF, in Italy on 10 January 1944. The extended-span wings were for high altitude operations.

A Seafire L Mk IIC with a low-altitude-rated Merlin 32 and four-blade Rotol propeller; it had the best low-altitude performance of any 1939-45 naval fighter. Seafires provided vital cover over the Salerno landings between 9 and 12 September 1943, flying continuous 20-aircraft beachhead patrols from dawn to dusk from nearby aircraft carriers, unlike the Sicily-based Spitfires which operated at extreme range. To enable Seafires to be launched rapidly from small escort carriers, especially at full load, Rocket Assisted Take-Off Gear (RATOG) was developed. Landing the Seafire — with its narrow undercarriage track — on carriers remained tricky — 114 were deployed for Salerno, and most of the 60 lost succumbed to landing accidents.

The Seafire Mk III was the first truly naval Spitfire, with stronger undercarriage to absorb punishing deck landings and folding wings to fit deck lifts for stowage in carrier hangars. It was first used operationally during Operation AVALANCHE, the Salerno landings. No. 809 Sqn, in HMS *Unicorn* and commanded by Major A. J. Wright, one of only two Royal Marines to fly Spitfires or Seafires operationally, operated from Paestum after the carriers of 'V' Force withdrew on 12 September 1943.

Navalised: Seafire Mk IBs, probably of No. 807 Sqn, RN. A total of 48 Mark IBs was converted from Spitfire Mk VCs by the addition of arrestor hooks, the first Seafire.

Spitfire Mk VIIIs of the 309th FS, 31st FG at Castle Volturno in February 1944 during the period of operations in support of the Allied landings at Anzio. The 309th returned its Spitfires to the RAF on 30 March 1944, re-equipping with P-51Ds.

Lt Hank Hughes with his 309th 31st Mark VIII *Audrey*, Castle Volturno, during the Anzio show.

Mark IXEs of No. 241 Sqn (RZ-R and U) over the Vesuvian plain in 1943. Capturing the airfields around Vesuvius, after the fall of Naples on 1 October 1943, and, a week later, the airfields on the Foggia Plain on the East Coast, gave Allied air power the bases it needed to prosecute the tactical war in Italy and to fly bomber missions from Italy to Germany.

.t Fawcett, this time with his 309th FS, 31st FG Mark VIII WZ-JJ, Castle Volturno, ?ebruary 1944.

ted. On 17 August 1943, Sicily was ompletely under Allied control.

There were three major results from he Sicily campaign, all of which had sig- ificant effects upon the future career of he Spitfire. First, the Allies had secured 1e launching-platform for the invasion f Italy; secondly, air power doctrines vere radically overhauled; and thirdly, icily's fall brought about Mussolini's ıll and the Italian armistice, thoroughly eakening the German position in Italy nd Europe.

Sicily provided a critical test for Allied ir power doctrines. They believed they ad relentlessly driven the Axis forces to the sea; in fact, the Axis withdrawal ad been pre-planned, well-executed and overed by massive AA. The Allies failed ɔ break the evacuation. Air operations concentrated upon ports and launching sites, but not the routes to them or the evacuation craft; 40 per cent of operations were conducted with heavy bombers, whereas battlefield operations by tactical aircraft were needed. As a result, effective air-to-ground support, begun in the Desert, was developed to bring tactical air power to bear where most needed, and the strategic use of air power and the commanders who used it were put in perspective. As Allied air superiority grew and the tactical lessons were applied, Spitfires were used less and less for fighter duties and increasingly in close support roles.

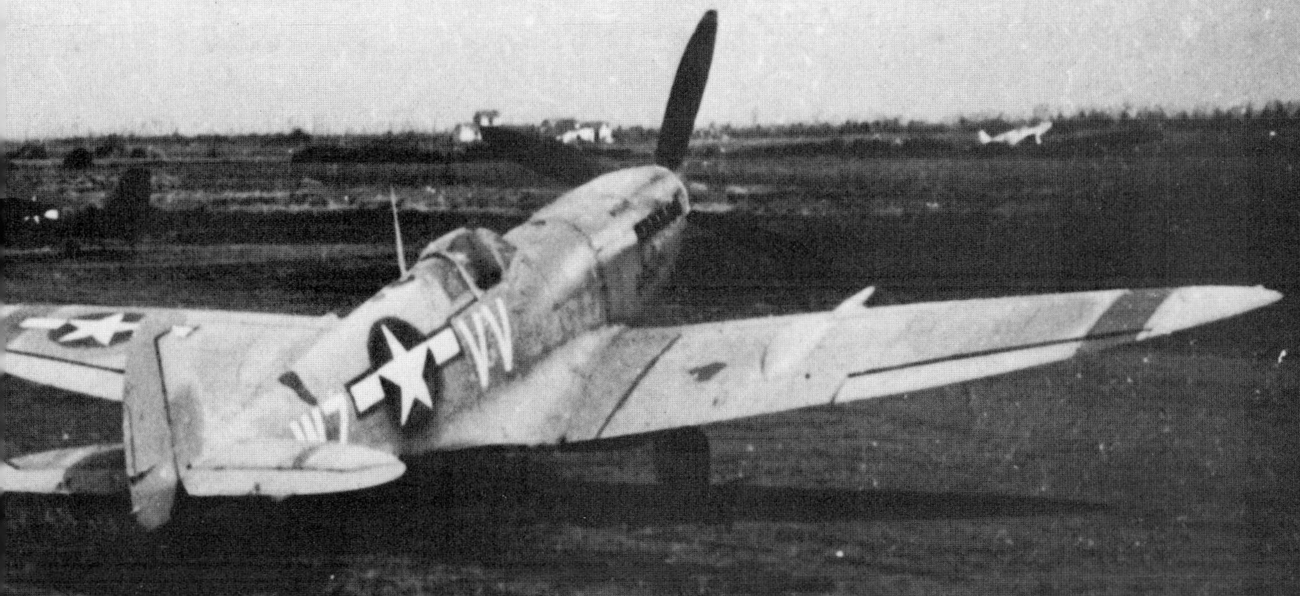

Invasions of France

On 15 November 1943, Fighter Command disbanded and was replaced by Air Defence Great Britain, but the majority of its squadrons were transferred to the Second Tactical Air Force (2TAF) completing its formation. The Marks IX and XI were to be the standard 2TAF fighter and PR Spitfires, to ease logistics and supply, and squadrons flying earlier marks were gradually re-equipped throughout 1943 and into early 1944. Half 2TAF's squadrons flew Spitfires — 34 — while Nos. 808, 885, 886 and 897 Squadrons, FAA, flew Seafires to cover the Normandy landings.

During the pre-invasion co-ordinated interdiction campaign, to destroy enemy communications, Spitfires carried out dive-bombing attacks, with two 250 lb. or one 500 lb. bombs, straffed and flew escorts and sweeps. As D-Day approached, all 2TAF's wings worked-up to full operational status and their Spitfires were all adapted to carry bombs and brought up to the latest modification standards. On 4 June, all aircraft were painted with wide black and white stripes around wings and fuselages for identification.

On D-Day, nine Spitfire squadrons provided the cover for the first landings. The Allied Expeditionary Air Force (AEAF) flew 14,674 sorties that day, but neither that day nor for the next two days did the German fighters appear although on 7 June Spitfires destroyed 12 Ju 88 bombers. AEAF air supremacy went unchallenged. On the ground, RAF Servicing Commandoes and Construction Wings opened new airstrips as the armies advanced, and, on 8 June, Wg Cmdr Johnson led No. 144 Wing into St. Croix-sur-Mer to become the first RAF unit to operate from French soil in fou[r] years. By 10 June, the Wing was based i[n] France.

On 9 June, Wg Cmdr Johnson sho[t] down his 27th victory, an Fw 190. Thre[e] Bf 109s destroyed towards the end o[f] June brought his confirmed victorie[s] level with the score of the RAF's top scoring fighter pilot, 'Sailor' Malan, wit[h] 30, now a resident of a PoW camp. O[n] 30 June, he shot down another enem[y] fighter. The Press feted him, but he di[d] not regard personal victory hunting a[s]

Wg Cmdr J. E. Johnson, his labrador Sally and Spitfire LF Mk IXE in Normandy.

Mark IX of No. 416 (City of Oshawa) Sqn, RCAF, No. 127 (RCAF) Wing, at Bazenville, France, 25 June 1944.

part of a fighter leader's job: getting maximum results from an entire wing was. Johnson's official score of 38 at the end of the war made him the RAF's official top-scoring fighter pilot of World War Two, but the high scores of the units he led is perhaps a better indication of his stature.

Luftwaffe reinforcement had begun to arrive and German and Allied fighters tangled daily. By the end of June, 30 Spitfire squadrons were based in Normandy. To combat the limestone dust of the area, tropical Vokes filters were hurriedly fitted. On 17 July, a Spitfire pilot made a significant contribution to the victory in Normandy, without knowing it. Sqn Ldr J. J. Le Roux, CO of No. 602 Squadron, was leading a 12-Spitfire armed recce in the area south of the front. After an abortive low-level combat with Bf 109s, Le Roux fired at targets of opportunity on a road. One was the staff car of Field Marshal Erwin Rommel, directing the German defence. It overturned. Injured, Rommel retired from the field.

On 15 August, when American forces landed in the South of France, Spitfires were in action in both northern and southern France. Among them were No. 451 (RAAF) Squadron's fighter-bombers, which flew five missions from Corsica that day. Called to readiness at 0400 hrs. and rolling down the flare-path at 0540, they first cleared an air lane off the French coast for transport aircraft and glider tugs. Later, they helped to cover the landing ships and convoys and the actual landings, then patrolled the beachhead until dusk.

Maintaining the 2TAF units in the field required a vast, tightly organised, highly mobile 'tail'. Groundcrew change the Merlin 66 of an LF Mark IXE of No. 442 (Caribou) Sqn, RCAF, in Normandy, 14 August 1944.

During the Battle of Caen, through August, Spitfires destroyed 400 tanks, 4,000 soft and armoured vehicles and 260 barges. Although the *Luftwaffe* still operated in large units over Germany, the fighting was defensive. Over France, the *Luftwaffe* operated in small units, minimising the number of aircraft involved. Moreover, the latest aircraft were difficult to keep together effectively in large formations. The RAF found, too, that large formations were unwieldy in the small scale combats that became the norm over France, and that, moreover, Spitfires were increasingly being deployed on tactical duties such as the Caen actions which demanded small, flexibly-employed units. Therefore, the three Canadian Spitfire Wings in France, Nos. 126, 127 and 144, No. 83 Group, which were operating in co-ordination, were re-formed into two Wings, Nos. 126 and 127, to operate separately, affording tactical flexibility.

Mobility was essential and the work was largely tactical. Nevertheless, throughout the campaign in North Eastern France, it was the primary function of the Spitfire units to protect the Typhoon fighter-bombers whenever the *Luftwaffe* was present, either by escorting them or by gaining local air superiority. By 7 August, the Germans were pinned into the 'Falaise Pocket' and put up a massive fighter umbrella to cover their withdrawal on the ground. There was a period of intense and costly fighting in the air. Nos. 126 and 127 Wings were based 25 miles from the Falaise Gap and straffed and bombed the retreating German ground forces as well as countering the *Luftwaffe*. The Spitfire Wings followed as the 21st Army Group moved up into Belgium, but some squadrons returned to the UK.

V-1s, V-2s and V-12s

Spitfires also formed the main ADGB force in early 1944, with 21 squadrons. ADGB's role was to maintain total air superiority over the UK, Channel and North Sea by flying offensive patrols to prevent *Luftwaffe* bombing and reconnaissance operations against the ports, supply bases, shipping and facilities vital to the success and forward impetus of the invasion. A very few *Luftwaffe* aircraft got through at night, to scant effect. By day, the UK was impenetrable to conventional aircraft.

However, on 13 June, the first V-1, a small, pilotless pulse-jet powered aircraft programmed to fly straight and level at 1,000-4,000ft, impacted at Swanscombe, Kent. Within 17 days, over 2,000 had been launched. Their target, however, was London, not the invasion assets, but many fell short. Air Marshal Sir Roderick Hill, commanding ADGB, implemented the anti-'Diver' measures: Operation CROSSBOW: a 480-strong barrage balloon barrier south of London; a formidable 400-gun AA battery chain on the North Downs linked to observation posts and radar stations —V-1s made excellent targets; and the first line of defence, fighters ranging in three patrol lines, Beachy Head-Dover, Dover-New Haven and, inland, Haywards Heath-Ashford.

Initially, six fighter squadrons were allocated, but soon other squadrons were being deployed as the attack intensified. The fastest low-level fighters were used: Spitfire LF Mks IX, XII and XIV, Tempest Mk Vs and Mustang Mks II and III by day, joined by the first jet, the Meteors of No. 616 Squadron, and Mosquitoes by night. Spitfires were not the fastest fighters on anti-'Diver' duties but were most numerous. In order to catch the 400 mph V-1s, the Spitfires' armour (there was no return fire!) and unnecessary equipment, including some guns, were taken out. On many, the paint was stripped and surfaces polished. Work done by RAE Farnborough increased, without serious modification, the low level speed of the Spitfire Mk XIV by 8 mph to 365 mph, while the Mark XI had a sea level speed of 346 mph, i standard condition, about 14 mph faste than the standard Mark IX; boosted the were faster. The Mark IXs were hard pu to catch the V-1s and had special boost and ran on 150 — instead of the usua 100 — octane fuel. Such modificatio wore engines out very rapidly.

Two Griffon IV-powered Mark XIIs, MB882 and MB858, of No. 41 Sqn in their element — down low — during the anti-'Diver' period. A total of 55 was converted from Mark VIII airframes, like this pair, while 45 were from Mark IXs. They were used defensively against 'hit-and-run' Fw 190s and Bf 109s in the south of England. Offensively, their tactics were for the brave. Flying very low, they invited German fighters to bounce them . . . no fighter matched them at low altitude.

riffon-pugnacity — the prototype Mark
II. Handling better than previous marks,
had an excellent low level performance —
p to 372 mph at 5,700ft. — and a
pectacular climb straight off the ground.

Of the Spitfires, the Griffon-powered Marks XII and the new LF Mark XIVE vere most successful. At Hawkinge, ome four miles from the Channel on he North Downs, Nos. 402 (Canadian) nd 350 (Belgian) Squadrons managed o keep 20 Mark XIVs on standing patols through the daylight hours. On 16 August, No. 402 Squadron's first success was a hat-trick practically over the airield. Ultimately, Nos. 41, 91, 130, 322 Dutch), 350 and 402 Squadrons flew Mark XIVs on anti-'Diver' patrols.

At the end of their service lives, the Mark XIIs were given a chance to shine n their true milieu — sprint work on the leck — and Nos. 41 and 91 Squadrons hot down many V-1s before their Mark XIIs were exchanged for Marks IXs and XIVs from September. No. 41 Squadron produced two V-1 aces, but No. 91 Squadron was graced by no less than 14, leaded by five with scores over ten: Sqn Ldr Knyaston had 17.

Attacking the V-1s was dangerous for he fighters. The target was smaller than fighter and needed a close attack. It was asy to become target-fixated and go in oo close, getting caught in the debris nvelope as the V-1 exploded. Captain Maridor of No. 41 Squadron did not become fixated. On 3 August, he was pressing home his attack from a safe distance on his eleventh V-1 when he realised it would fall on a military hospital. He accelerated and fired at point blank range. He stood no chance.

On 23 June, a Spitfire Mk XIV pilot, having noted that near-miss AA fire upset the V-1s gyroscopic controls, flew alongside a V-1, carefully slid his wingtip under the V-1's wingtip, felt the suction between the two wingtips, banked slightly, and the V-1 tipped, its gyroscope tumbled and it fell on to its back and powered into the ground. From then on, killing V-1s became safer if not easier.

On 17 July, the AA gun chain was moved to the coast and soon was taking out more V-1s than the fighters. The Allies overran the Northern France V-1 launch sites that had not been destroyed by bombers by 2 September; a few were subsequently launched from Holland or from bombers, but the battle was over. Of 3,638 V-1s destroyed, balloons claimed 278, guns 1,560 but the fighters claimed the majority, 1,900.

However, on 8 September, a more sinister threat, uncounterable by direct means, had emerged — the V-2 ballistic rocket, launched from sites in Holland, initiating truly modern total war. An anti-V-2 force was rapidly brought together, comprising Nos. 229, 453 (RAAF), 602 and 603 Squadrons, based in the UK, flying Spitfire LF Mk XVIE fighter-bombers equipped with either two 250 lb. bombs under wing or a 500 lb. under the belly. On 3 December, Nos. 602 and 453 Squadrons bombed flats at Maagsche Bosche which housed the V-2 troops' HQ. Suspected stores and launch sites were dive-bombed and the Hotel Promenade at The Hague was also attacked. One pilot attacked a launch site at the moment a V-2 was lifting off, and put in a long but unsuccessful cannon burst! On Christmas Eve, Nos. 229, 453 and 602 Squadrons bombed-up in England with two 250 lb. *and* one 500 lb. bombs, refuelled in Belgium, and made heavy attacks on the launching sites. The V-2 attacks continued at a reduced rate until 27 March 1945.

The End in Europe

On 9 September 1944, another of Germany's revolutionary weapons was first sighted, the Me 262 jet fighter/fighter-bomber, by Spitfires of No. 411 Squadron, but it easily evaded their attack. Me 262s made regular hit and run attacks on Allied airfields in Belgium. Standing patrols brought little protection. Although five No. 401 Squadron Sptifires destroyed the first in combat near Nijmegen on 5 October, just as the Fw 190A had outclassed the Spitfire Mk V, so the Me 262 outclassed the Spitfire Mks IX and XVI, and every fighter in the Allied inventory. The speed differential was so great that no development of the Spitfire could hope to match the Me 262, but if it were drawn into a maneouvring fight the Spitfire had a decided advantage, as Flt Lt J. J. Boyle of No. 411 Squadron, No. 127 Wing proved on Christmas Day 1944, when he brought down two.

A number of Spitfire squadrons in 2TAF, including those in No. 127 Wing, re-equipped with the Spitfire LF Mk XVIE, essentially a Mark IX airframe with a US Packard-built Merlin 266, optimised for tactical work. Several of these aircraft and a number of pilots were lost on this Mark when the engine seized in flight because of faulty quality control during manufacturing. The con rods were machine-polished, which 'drew' the surface metal, concealing cracks from standard crack detection tests; under boost, the con rods broke. Repeated demands by squadrons for their Mark IXs met no response. Fred W. Town, recalled his experience of the Mark XVI with some disaffection:

The Spitfire had then the reputation of th[e] superior fighter and it did have the rate of tur[n] and the guns to prove it. On 4 December, N[o.] 403 Squadron (RCAF), of No. 127 Wing, re[-]equipped with Spitfire Mk XVIEs from th[e] Mark IX. We hated to see the Spit IX leave fo[r] it was the ultimate in Spitfires with the Rolls[-]Royce Merlin engine and elliptical wings[.] Now we were given the Mark XVI, and, ult[i]mately it was a wonder it ever got off th[e] ground. We could carry two 500 lb. bomb[s] and a slipper tank, but the Packard Merlin di[d] not compare with the Rolls-Royce-buil[t] Merlin as there were many times when th[e] crankcase and pistons could not handle th[e] boost. The three feet clipped off the wing spa[n] and taking-off from a grass field or a quickl[y] laid strip certainly did not help lift at take-of[f].

The Griffon-engined Spitfire Mk XIV[,] based upon the Mark VIII, was also ap[-]pearing in Belgium with No. 125 Wing[,] comprising Nos. 41, 130 and 610 Squad[-]rons.

In Italy, the harsh winter of 1944-4[5] forced a lull in the intense close ai[r] support activity, halting the Allie[d] advance on the River Senio on the Eas[t] Coast. Torrential rain turned airfield[s] into swamps. Often, cloud was too lo[w] for dive-bombing and air support wa[s] frequently impossible in any form. Th[e] highly-developed efficiency of th[e] ground-to-air co-operation could, how[-]ever, overcome some of the weather. O[n]

F Mark XIVE of No. 402 (Winnipeg Bears) Sqn, RCAF, seen at Heesch, the Netherlands, 4 March 1945. The Mark XIV was developed from the Merlin-powered Mark VIII by installing a two-stage supercharger Griffon which required a five-blade propeller to use the power, a large fin to counteract the torque and larger radiators to cool it. The 'bubble' canopy introduced on late production aircraft considerably improved rearwards vision. With clipped wings and slipper fuel tank, the FR (fighter reconnaissance) variant had an oblique F.24 camera in the rear fuselage.

one occasion, with the cloudbase down to 1,000ft. and support urgently needed, Dundas, No. 244 Wing's CO, requested that the artillery put down smoke shells where the support was needed, then he took a squadron of Spitfires and, guided by forward observation posts, his twelve Spitfires swept down through the clag in line abreast and, spotting the smoke, straffed the area effectively.

In January, three months of intensive interdiction activity began in preparation for the re-opening of the assault on the Gothic Line. There had been sporadic air combat in the months since Rome fell but in the last months of the war in Italy, Spitfires concentrated upon ground support, flying from first light to dusk, bombing and straffing. The assault, supported by massive aerial and artillery bombardments, engaging both the DAF and US 15th TAF, reopened in April. The long interdiction campaign had been fully successful; the German defence, shallow and frameless, crumpled.

The crossing of the River Senio in April brought the last battle. A number of Spitfire wings were engaged in ground support, which included straffing the north bank of the narrow Senio with Allied troops on the south bank only 20 yards away! Bombing and straffing operations were continuous throughout the daylight hours. No. 601 Squadron flew 1,082 hours during April, becoming the first squadron in Italy to exceed 1,000 operational hours in any month, while it also established a record in destroying motor transport, motor cycles, barges and horse carts.

Groundcrew bodily keep the tail down, while armourer harmonise the guns of a No. 417 (City of Windsor) Sqn, RCAF, Mark VIII at Al Fano, Italy, 16 December 1944. From June 1944 — when Rome fell and most *Luftwaffe* units were moved to Northern France to counter the Normandy invasion — Spitfires in Italy went over entirely to close support operations, with great effect.

In North West Europe, on 1 January 1945, the *Luftwaffe* launched Operation BODENPLATTE on the 2TAF, intended to be a decisive blow, using maximum resources. It wrought destruction at the 2TAF's bases destroying 144 aircraft, including many Spitfires, and killing many airmen, but the *Luftwaffe* could not follow it up. The Allies were able to replace their losses quickly, but the *Luftwaffe* could not recover from the grievous losses of experienced pilots it had suffered. Despite this, the *Luftwaffe* asked no mercy and, even when restricted to an untenable area, flew constantly until the last day of the war.

The tactical use of air power was honed to perfection during the onslaught on Germany. No. 83 Group — in which Spitfires predominated — during its operational existence alone accounted for 1,500 enemy aircraft, 700 locomotives, 4,000 railway wagons, 10,000 soft vehicles, numerous tanks and 300 vessels and barges, although it lost 900 pilots and 1,200 aircraft. In the face of overwhelming Allied air power, defeated on the ground in the east and west and precipitated by Hitler's suicide, Germany collapsed, although the fighting went on to the bitter end. Fred W. Town, recalled his last days of war:

> One of the Spitfires that I flew, LF Mark XVIE TB752, is now preserved at Rochester airport, Kent. I flew TB752 four times in the last three days of the war, while mine, KH-T, was being serviced. TB752 was KH-Z, for Sqn. Ldr. Zary, now deceased. On one trip in TB752, having released two 500 lb. bombs, I chose to stay low to avoid flak and after a few miles, climb for home, but an He 111 with an upper gun turret had me trapped so, with ammunition still available, I was in position easily to strike and destroy the aircraft. The third flight was bombing Heilingenhafen 'drome which housed some 100 aircraft. There were many ships nearby. I was able to see five bombs alight — two on the runway, two among aircraft and one among the hangars. My own two bombs must have caused damage as the targets were so accessible with very little flak. The last trip with this aircraft was dive-bombing a 15,000 ton ship near Kiel, resulting in one direct hit and one near miss — fortunately there was little flak. Her very last trip was on 5 May, the day that Germany surrendered unconditionally at 0800 hrs., when our Squadron's 12 aircraft — I flew KH-T — escorted 14 Dakotas with VIPs to the Peace Conference.

The End in the East

The first Spitfire Mk VCs arrived in India in September 1943 and in succession Nos. 607, 615 and 136 Squadrons re-equipped, superseding the Mohawks and Hurricanes which were easy meat for *Zeroes*. The Spitfires' arrival transformed South East Asia Command's capability and within months they had won air superiority. In January 1944, two squadrons became operational with Mark VIIIs which were far superior to any Japanese fighter in the theatre.

The air war in the Burma theatre was very different to any other. It was dominated by the weather, jungles, hill ranges, rivers and mountains and relied heavily upon air transport. Conditions were often atrocious. It was a major part of the task of the fighters to win air superiority to allow the transports to function. This was demonstrated in April 1944 when the great battle of Imphal developed. The Japanese were beleaguering Indian and British troops who could only be supplied by air. Spitfires won air superiority rapidly and unquestionably and allowed the Allied troops to hold the perimeter.

Thereafter, with negligible resistance in the air, Spitfires increasingly went over to ground attack work under army control, flying in pairs. However, they were not ideal for such work and it became policy to re-equip RAF fighter units in SEAC with the Thunderbolt which was superior as a fighter-bomber.

Spitfires, however, were to be engaged in the invasion of Malaya, Operation ZIPPER, when they were to be flown off to advance bases from carriers. However, the war ended while plans were well advanced and only No. 17 Squadron flew off a carrier, HMS *Trumpeter*, bound for Kelanang in September 1945 to take up occupation duties.

Further to the east, the Spitfire Mk VIII Squadrons of the RAAF's No. 80 Wing were engaged in ground support duties in the New Guinea area. However, debarred by the Americans from operating in the Philippines where MacArthur wanted all the glory, they were squandered upon secondary targets at extreme range, resulting in heavy attrition and loss of experienced pilots for very little gain. In 1945, this led to several senior Australian operational commanders resigning their commissions, including 'Killer' Caldwell, in protest at the waste of life and potential. As the war in the South West Pacific dragged to a close, there was little that the Australian High Command could do except order a bureaucratic enquiry to cover the mess up.

Once *Tirpitz* had been sunk and the Allies were established on European soil, the Royal Navy was free to send capital units to fight in the Far East. British Fleet carriers were assembled into Task Force 57 to join the US Navy in the final onslaught on Japan. *Indefatigable* embarked Nos. 887 and 894 Squadrons and *Implacable* Nos. 801 and 880 Squadrons of Seafire L Mk IIIs. TF 57's other Fleet carriers embarked US aircraft types, easier to replace and supply via the US Navy. However, Seafire L Mk IIIs also served in Far East waters with Nos. 807, 809 and 879 Squadrons aboard the light carriers *Hunter, Stalker* and *Attacker* respectively.

In March and April 1945, *Indefatigable*'s Seafires were active during operations off Okinawa. Although their short range restricted them to flying Comba[t] Air Patrols (CAPs) over the Fleet, the[y] shot down a number of Japanese aircra[ft] which attacked the Fleet. However, th[e] deck landing accident rate was 7 per cent too high for operations. After fitting stronger tyres, replacing some Deck Landing Control Officers and replacing tired pilots with fresher pilots from No 899 Squadron in the Middle East, th[e] rate fell to 2 per cent.

In July, *Indefatigable* joined TF 5[7] which reached its strike area on 16 July 1945, *Implacable*'s and *Indefatigable*'s Seafires and Fireflies flying over the US

Pranged No. 681 (PR) Sqn, PR Mark XI a[t] Alipore in overall PR Blue but for white SEAC bands on wings, fin and tailplane. The first Spitfires in India-Burma were No 681 (PR) Sqn's PR Mark IVs. Accurate, up to date tactical intelligence was essential in Burma where distances, terrain and weather made communications and locating the enemy difficult. The PR Mark XI was developed from the Mark IX fighter and was fully tropicalised. The availability of a ladder exactly the right height to reach the tail of an up-ended Spitfire is interesting . .

No. 607 Sqn pilots and Mark VIIIs at a water-ridden Kalaywa (Toungoo) in mid-1945. Small blue and white roundels were adopted to avoid confusion with the Japanese red rising sun insignia.

Third Fleet to signal their arrival. The Seafires were now fitted with P-40 belly drop tanks which, besides improving their deck-landing characteristics compared to the slipper-type, extended their radius of action. More much potent as naval fighters, they could be used offensively over the Japanese islands. On 17 July, the Seafires flew their first sweep from a position just 170 miles east-north-east of Japan. For the next two and a half weeks, they flew escort and strike missions over Japan, but, on 3 August, a secret US Fleet order suspended all air operations until a 'special operation' had been undertaken. Three days later, on 6 August, Hiroshima was atom-bombed.

'Killer' Caldwell, top-scoring Australian fighter pilot, and his victory tally — Germans, Italians and Japanese.

Mark VIII of Wg Cmdr Caldwell's No. 80 Wing, operating from Morotai, with bomb carriage shackles and slipper tank for the long-range ground attack role. The cartoon is ironic in view of the official furor over the supposed 'grog-smuggling' role of RAAF Spitfires to US forces!

Old Soldiers . . .

Wartime Spitfire orders were tapered out gradually to avoid large scale unemployment, but the operational use of Spitfires was abruptly diminished. The main Merlin-Spitfire in RAF use post-war was the LF.16E, but all Mark IXs were made available for export within days of the war ending. At home, jet re-equipment was underway by 1946, and by the end of the year there were only two front-line Spitfire fighter squadrons in the UK, No. 63 at Middle Wallop and No. 41 at Wittering in Nos. 11 and 12 Groups respectively, and No. 41 Squadron was in the process of re-equipping with DH Hornets. No. 63 Squadron finally re-equipped in April 1948. However, No. 541 (PR) Squadron flew PR.19s in the UK, doing photo survey work, until 1951, the last home-based Spitfire PR unit.

On the Continent, the Spitfire squadrons in Italy had disbanded within a few months of the war's end, although some were transferred for occupation duties in Austria. In British Air Forces of Occupation in Germany, many squadrons, particularly those flying Mark IXs and Allied and Commonwealth units operating with the RAF, were disbanded in 1945-46. No. 83 Group soon embraced all Spitfire units in BAFO Germany, but gradually replaced them, the last fighter squadron, No. 80, leaving in July 1949 for the Far East, although No. 2 (PR) Squadron flew FR.14Es and PR.19s, doing useful survey work until disbanding in 1951.

The Spitfire was granted a new lease when, in June 1946, the AuxAF was re-constituted. On 16 December 1947, the AuxAF's prominent contributions during the war were recognised by the prefix 'Royal'. The 20 squadrons, as pre-war, were identified with localities and counties. Seven flew B/NF Mosquitoes, and 13 flew Spitfires in the day-fighter role. Although some squadrons flew LF.16Es, Griffon-Spitfires were used in some numbers between 1946 and 1951. Some squadrons were initially equipped with FR.14Es, some of which re-equipped with F.21s in 1947, but the F.22 became the main fighter in 1948-49, Nos. 603, 607, 611 and 613 Squadrons still flying them during the reserve call-up to active service in 1951, before re-equipping with jets.

Miscellaneous home units flew Spitfires for some years after the war, such as Group communications squadrons while the Control Gunnery School had 12 LF.16Es. In 1949, analysis of Spitfire accidents during day-time training and communications flights revealed 125 in 15 months. The later, heavier marks were not suitable for training as had been the lightly loaded Marks I and II. By 1950, in the UK the chief users of Spitfires were Anti-Aircraft Co-Operation Squadrons, LF.16Es and F.21s, until as late as 1954.

At the end of the war, the Fleet Air Arm had rapidly contracted to less than a twentieth of its peak size within a few months, largely through the demobilisation of the RNVR Air Branch and the disposal of the US lend-lease types equipping the majority of squadrons. However, without the US aircraft, there was

Royal Navy Seafire F.15 (Griffon IV) PR499 in the USA. Basically a Spitfire Mk XII with Seafire Mk III folding wings and naval equipment, the F.15 entered RN service as World War Two ended. It was developed into the F.17, with a 'bubble' canopy and stronger undercarriage.

No. 273 Sqn FR Mark XIVE over Malaya in late 1945.

The standard Middle and Far East post-war RAF Spitfire was the F/FR.18, produced in both fighter and (illustrated, June 1945) fighter reconnaissance variants. (RAF mark numbers were Arabic from 1946.) The ultimate development of the basic Spitfire airframe, it resembled the late production Mark XIV and used the same Griffon variants, but had a new, stronger wing spar.

An Israeli LF.16E, one of at least 30 LF.9s and 16s obtained by Israel from Czechoslovakia whose Soviet-dominated Government disapproved of using 'Imperialist' equipment in December 1946. (The 76 transferred from the RAF to the reconstituted Czech Air Force in 1945-46 included those of the RAF Czech Wing.) They were ferried out to Israel via Yugoslavia in 1946. The Israelis used them in action against Egyptian forces, which also flew LF.9Es. Superseded by jets, the surviving Israeli Spitfires were refurbished by Bedek at Lod Airport and sold to the Union of Burma in 1954-55.

The definitive PR Spitfire, the Griffon-engined PR.19, developed from the Mark XIV. This example (February 1945), has a 170 Imp Gallon slipper tank which, with internal fuel, gave a 1,550-mile range — the greatest of any Spitfire.

a place for Seafires and the F.15, entering operational service just as the war ended, served until 1951 with RN and RNVR units, latterly only ashore, alongside a refined development, the F.17. The later Seafires, F.45 and F.46 were not very successful, although the last, the F.47, served with Nos. 800 and 804 Squadrons, seeing action with the former in Malaya and Korea in 1949-50, and becoming the last operational RN Seafire.

Many demobilised RNVR personnel joined the AuxAF, but under a 1946 Admiralty scheme, four RNVR air squadrons were formed by late August 1947 in the best recruitment areas, to

Royal Thai Air Force F.14E at Bangkok, March 1956. It has underwing bomb shackles and points for 5in. rockets (HVAR). Thailand bought 30 reconditioned ex-RAF F/FR.14Es from Far East stocks.

retain at least some of this huge reserve: No. 1830 at Abbotsinch with Firefly FR.1s, and Nos. 1831, 1832 and 1833 at Stretton, Culham and Bramcote with Seafire F.15s and F.17s. They had initial establishments of 12 pilots aged 20 to 28 and 14 to 18 aircraft. Like the RAuxAF, flying practice took place at weekends at their bases and during 14-day annual summer camps, including several embarked in carriers. Nos. 1831 and 1832 re-equipped with Hawker Sea Fury FB.11s in August 1951. In August 1953, No. 1833 Squadron discarded its F.17s but received FR.47s, embarking in *Triumph* that summer, the last RN Seafire unit to be embarked. No. 1833 Squadron finally re-equipped with Sea Furies, ending the Seafire's RN service, on 15 February 1954.

Overseas, RAF Spitfires served in front line roles until the early 1950s, standardised upon the tropical F/FR.18. In the Middle East, the FR.18 equipped No. 32 Squadron until 1947 and No. 208 Squadron until 1951, operating in Palestine, and both units tangled with Egyptian and Israeli Spitfire 9s and 16s. In the Far East, the F/FR.18 became the RAF's standard front-line fighter post-war, replacing F/FR.14Es and Thunderbolts in India and Malaya. Spitfire squadrons were stationed at Penang, Singapore and Hong Kong, and because of the nationalistic fervour in the regained colonies, they were kept at a high state of readiness for some months after the war ended. The first crisis came within months.

Free of the Japanese, revolutionaries in the Netherlands East Indies proclaimed the independent state of Indonesia, and began to arm with ex-Japanese weapons and training aircraft. There were several clashes with rebels, but a serious battle began at Surabaya in November 1945. RAF support was called in. No. 155 Squadron's Spitfire Mk VIIIs, joined later by Mk XIVs, No. 60 Squadron's Thunderbolts and four Mosquito squadrons were deployed. They began armed reconnaissance and pin-point strikes. Notably, Mosquitoes silenced three rebel-controlled propaganda radio stations and Spitfires and Thunderbolts a fourth. From then until British operations against the insurgents were completed in March 1946, the Spitfires, Thunderbolts and Mosquitoes supported the ground forces, attacking road blocks and strongpoints directed by Forward Air Control (FAC) Austers, without once being hit. This was the last operational use of Merlin-Spitfires by the RAF.

The British forces were replaced by Dutch forces during 1946, equipped

with modern weapons and aircraft, including 20 Spitfire Mk IXs, against which the rebels had no comparisons. The rebel aircraft were rapidly destroyed on the ground or shot down.

Order seemed to have been restored in the Far East by mid-1947 and the Far East Air Force Command halved the strengths of the Singapore Spitfire squadrons, Nos. 28 and 60, to eight FR.18s each. However, in May 1948, with organised and numerous communist insurgents active in the country, the Malay Government declared a state of emergency. British infantry mobilised up country, and, in July, Nos. 28 and 60 Squadrons moved to Kuala Lumpur. No. 45 (Beaufighter) Squadron joined them from Ceylon. A counter-insurgency campaign, code-named FIREDOG, began, Spitfires carrying out attacks on suspected insurgent camps, straffing jungle clearings and making pin-point gun and rocket attacks. Materially of little value, the effect on the insurgents' morale was keen. In May 1949, No. 28 Squadron moved to Hong Kong but between October 1949 and February 1950, the Seafire FR.47s of No.800 Squadron, RN, flew strikes in Malaya from Sembawang, Singapore.

The Spitfire PR.19s and Mosquito PR.34s of No. 81 Squadron flew target intelligence and photo-reconnaissance sorties during FIREDOG. As low-level flights would alert insurgents, sorties were flown at normal PR heights, 16,000-17,000ft. Camps located on photographs were hit by air strikes of various types. In one type, the Army 'swept' an area, forcing insurgents into the grasslands, the *lalang*, where Spitfires straffed them.

During 1950, more effective, later-generation strike aircraft began operations in Malaya. No. 60 Squadron was the last RAF Squadron to use the Spitfire as a fighter, making the last operational attack on 1 January 1951 on a suspected insurgent base in the Kota Tinggi area of Johore. No. 81 Squadron continued to operate PR.19s over Malaya.

In 1948, Britain became alarmed about the consequences of a communist victory in the Chinese civil war upon the territories, including Hong Kong, held under lease from China. If hostilities broke out, the UN would assist, but Britain first decided upon a show of strength. An exercise in 1948 demonstrated how effectively Hong Kong could be reinforced. Six Spitfires and a Mosquito, followed by an ASR Sunderland and a Dakota with their equipment and ground crew, flew from Singapore to Hong Kong via Kuching, Sarawak and Clark Field, Philippines.

In 1949, as the Communists' victory became certain, a reinforcement plan was implemented. The Spitfire FR.18 squadrons were brought up to war strength, 16 aircraft. It was planned that Kai Tak, Hong Kong, would operate three Spitfire squadrons, but only No. 28 Squadron, transferred from Malaysia, and No. 80 Squadron were sent. No. 80 Squadron was withdrawn from Gutersloh, Germany, on 26 July 1949 and flew its F.24s via Manston to Renfrew where Airworks dismantled them, for shipping by carrier direct to Hong Kong. HMS *Unicorn* was despatched to Hong Kong, her Carrier Air Group including Seafires. It was a far cry from 1942, when Britain had been unable to reinforce Hong Kong against the Japanese threat. The display of strength had the desired effect.

However, much more serious communist expansion in the Far East distracted the UN and China from Hong Kong, when, in June 1950, North Korean communist forces swept into South Korea. The UN mobilised in defence of South Korea. *Triumph*, the only British carrier in Far East waters, was placed at the disposal of the UN Task Force Com

The ultimate Spitfires, the F.21, 22 and (illustrated) 24, differed in detail but late F.22s and all 27 F.24s had a larger fin/rudder. They formed the basis for the last Seafires, the FR/F.45, 46 and 47, respectively. The first radical departures from the basic airframe, they had entirely new, less distinctly elliptical, laminar flow wings and four-cannon armament. The need to maintain wartime production impetus ousted introducing radically new production jigs and the 'Super Spitfire' arrived too late, usurped by the jets.

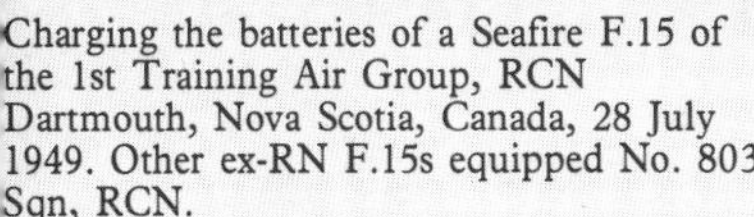

Charging the batteries of a Seafire F.15 of the 1st Training Air Group, RCN Dartmouth, Nova Scotia, Canada, 28 July 1949. Other ex-RN F.15s equipped No. 803 Sqn, RCN.

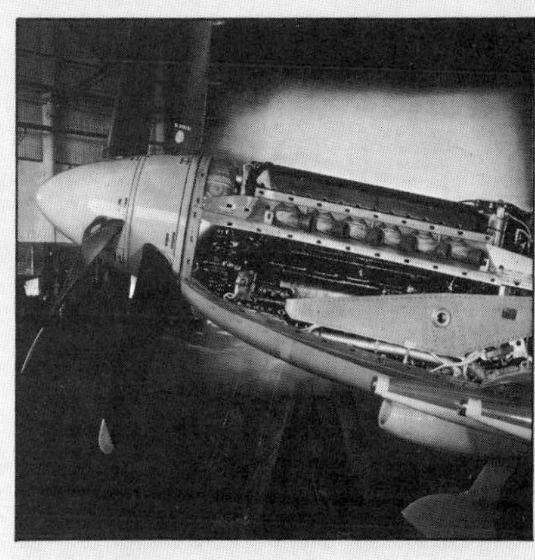

Unrecongisable as an evolution from Mitchell's original Type 300 Spitfire, a Seafire F.46's power plant — the 2,045 hp Griffon 85 driving a six-blade contra-rotating airscrew. Serviceability problems led to the Mark's early demise.

mander, General MacArthur, joining USS *Valley Forge*. Deprived of land bases in South Korea, the UN's only means of striking the enemy was by carrier aircraft. The 13th Carrier Air Group aboard *Triumph* flew Firefly FR.1s and Seafire FR.47s, the latter of No. 800 Squadron which embarked from Sembawang, Singapore in autumn 1950. The RAF was not involved in the Korean War; nor were Spitfires: they were obsolete in modern war.

Triumph's Seafires and Fireflies were active during the early period of the Korean war, flying 895 sorties; strikes against road, rail and port installations in the interdiction role, attacking junks suspected of being minelayers, and flying CAPs over ships operating off the Korean coast. The Seafires flew 115 ground attack operations and 245 offensive fighter patrols. These were the last offensive operations flown by Royal Navy Seafires. In late 1950, *Theseus* arrived from the UK with the Sea Fury FB.11s and Fireflies of the 17th CAG to replace *Triumph*. When No. 800 Squadron relinquished its Seafires in early 1951, the Seafire left front-line RN service.

In January 1952, No. 80 Squadron replaced its Spitfire F.24s with DH Hornets, and the Spitfire fighter left RAF front-line service. By then, only six serviceable RAF Spitfires remained in the Far East, five PR.19s operated with eight PR.34 and one T.3 Mosquitoes by No. 81 Squadron, and one F.18 of the Seletar Station Flight. On 1 April 1954, PR.19 PS888 made the last operational photographic reconnaissance sortie by an RAF Spitfire, over Malaya. The same day, the PR.19s were struck off charge, the last operational RAF PR Spitfires.

Three PR.19s were the last Spitfires employed actively in the UK. They served with the THUM Unit, formed at Hooton Park in 1951, and based at RAF Woodvale. Under Air Ministry contract with Short Brothers and Harland Ltd., they did meteorological work for the Central Forecasting Office at Dunstable. Specially equipped as 'MET.19s' by AGT of Gatwick, they flew 4,000 sorties until replaced by Mosquitoes in June 1957. The two survivors, PM631 and PS853, the latter in the hands of Group Captain 'Johnnie' Johnson, were flown to RAF Biggin Hill to become founder members of the Battle of Britain Memorial Flight, with whom they still fly.

RAF (Retd)

The Battle of Britain Memorial Flight was founded in 1957 with the purposes of preserving in flying condition the most evocative British aircraft of World War Two, and of promoting the RAF and encouraging recruiting. This it has done superbly, enthralling the public with displays by its Spitfires, Hurricanes and Lancaster, and, despite the strictly controlled flying hours available per aircraft, packing in some 80 airshows, exhibitions and fly-pasts each year.

Since its formation, the Flight has operated almost a score of historic aircraft. Currently, it has a very healthy complement of four Spitfires, one of which fought in the Battle of Britain, two Hurricanes and a unique Lancaster Mk I. Each of the Spitfires and Hurricanes — past and present — has taken a different road to survival. Chance has played a fair part, but their survival has depended in several cases upon the dedication of individuals and teams.

The Flight's flying crews are all serving RAF officers and NCOs. There is no shortage of volunteers, but the selection process is rigorous and unique. A pilot has to adapt rapidly from tricycle undercarriage aircraft powered by smooth jet engines to 'tail-draggers' with complicated piston engines. A large number of aircrew are needed because they rotate duties with the Flight with their duties on operational squadrons which take first place.

The ground crews are all on th Flight's permanent establishment. Al are experts in their fields, many wit specialised skills which have been obso lete in the RAF for over two decades Their task is not easy. Spares for th

The first US-based Spitfire on the US civil register, N7929A, an F.24, destroyed in a fatal crash after oxygen system failure at high altitude.

Spitfire FR.14E GF-GMZ, ex-TZ138 which served in Canada for winterisation tests in winter 1946-47. Subsequently, it was a National Air Racer in the USA. It is seen at Cleveland Main Airport on 9 June 1949 during the Cleveland Air Races when it came third in the Tinnermann Trophy Race at 359.565 mph. Allowed to langour at Perry Airport, Hollywood, it was restored in Minneapolis in 1960. Owned by Len Tanner, it was loaned to the Bradley Air Museum, Windsor Locks, Connecticut.

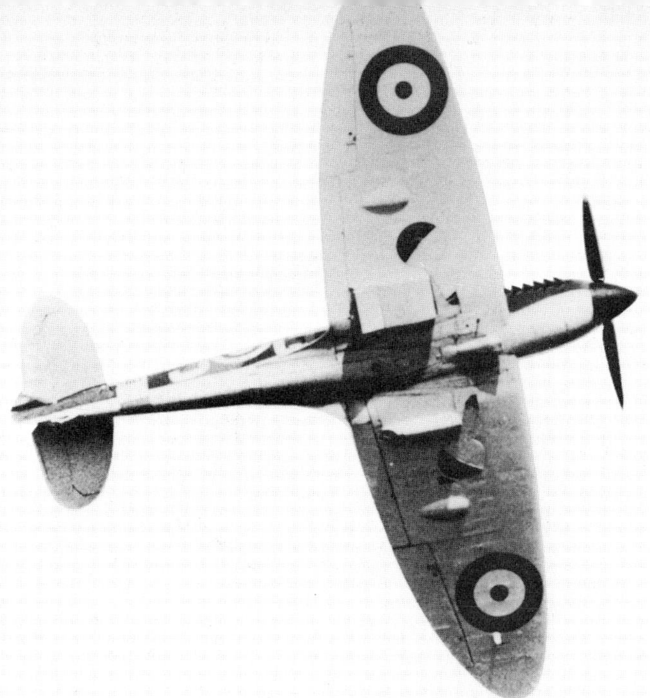

Probably the most photographed Spitfire of all time, Mark VB AB910, a member of the Battle of Britain Memorial Flight since 1965. It was built at Castle Bromwich in 1941 and served with Nos. 222 (Natal), 130 (Punjab), 133 (Eagle), 242 (Canadian), 416 (Canadian), 402 (Canadian) and 527 Sqns — a career that makes it a fitting member of the Flight. In 1947, it was sold to a private owner for racing but was bought in 1951 by Vickers-Armstrong and rebuilt for displays.

45-year old aircraft are hard to find, and the ground crews are constantly seeking new sources. Overhaul, servicing and flight schedules are strictly obeyed, and flying hours, manoeuvres and 'g' imposed in displays are stringently monitored. Regular and meticulous maintenance is essential, and periodic rebuilds are undertaken to ensure that aircraft are not only fully airworthy but able to survive for many years because the Flight intends to keep its aircraft airworthy and performing at displays well into the twenty-first century.

It is not surprising that many Spitfires have survived the scrapper's torch to become airfield guardians, memorials, museum exhibits, private air display aircraft and 'film stars'. The Spitfire has become one of the most sought after of airworthy veteran aircraft. Restorers and collectors have spent large sums of money upon finding, acquiring and extensively rebuilding examples, some from mere hulks, to flying condition or museum standards — and a Spitfire recently changed hands for half a million pounds.

Spitfire LF.16E SL721 now owned by Woodson K. Wood in the USA. This aircraft, built at Castle Bromwich, served on the Fighter Command Communications Sqn in 1946 and was used by Air Marshal Sir James Robb, hence JM-R, then on the Meterological Communication Sqn in early 1948 and with No. 31 Sqn from late 1948. Struck off RAF charge in 1951, it was displayed at the Beaulieu Motor Museum, Hants, until 1969 when it was sold to Bill Ross of Chicago, Illinois, and restored to flying condition, registered N8R. Sold to avid Spitfire collector Doug Arnold, it was registered G-BAUP to Warbirds of Great Britain Ltd, Blackbushe Airport, Camberley, Surrey, from 4 April 1973 until 21 July 1977 when it was sold to Wood.

LF Mark VIII MT719 immaculately restored in the early 1980s in its original No. 17 Sqn SEAC markings, YB-J, flown by D. K. Healey. It visited No. 17 Sqn at Bruggen, RAF Germany, in 1984, which operates Jaguar GR.1s in a not dissimilar tactical role to its Burma forebear. MT719 was found in a dump in India, having been operated by the Indian Air Force.

But, for the pilots and thousands of others, no monetary value could summarise the stirring of associations evoked by that strange name — Spitfire! Perhaps the last words should be left to the different perspectives of a young man who flew the aeroplane in combat, Fred W. Town, and a Supermarine test pilot, Don Robertson.

It was a thrilling experience to fly a Spitfire for it was a basic machine with no fancy frills. When you sat on your parachute, ducked your head to close the hood, your body was touching both sides of the cockpit, your maps were handy in your flight boots, and your radio as part of your helmet was automatically part of necessary equipment that made the Spitfire an extension of your own body. The feet and hands created the beautiful bird-like movement and two cannons and four machine guns could create destruction with the press of the thumb.

Fred W. Town

There must be many men alive today who remember their first sight of a Spitfire, poised rather delicately on a somewhat spidery looking undercarriage but nevertheless with a purposeful lean and hungry look bristling with guns. Boarding the aeroplane with a bulky parachute strapped to his back was a clumsy job for the pilot but with the left foot on the trailing edge of the wing and the right hand on the back edge of the tiny door it was easy to step forward on the wing root, turn half round and swing the right leg into the cockpit, stepping on the cup-shaped metal seat mounting for the parachute. Holding on to the top of the windscreen with both hands, a slither down and with feet sliding forward he would settle comfortably into his familiar seat with everything in reach. Holding the umbilical radio and oxygen leads clear, the shoulder straps would be handed to him followed by the lap straps to fasten to the central pin with the elementary quick release gear. Old sweats would slump down in their seat while adjusting the tension so that on sitting up normally the straps were really tight with the backside pressed hard into the seat; to shoot well it was essential to feel like part of the aeroplane with the shoulders taking the recoil of the guns. Then followed the chores: plugging in the R.T, inserting the bayonet attachment for the oxygen and a quick run over the petrol gauge, gun safety catch and setting the altimeter to zero, finally releasing the lock on the Sutton harness to hold it back for take-off.

Starting from cold about five strokes of the Ki-gas primer should be enough, throttle open a crack and with trolley accumulator plugged in, left hand on the throttle right on the ignition switches ('up to the sky to fly'), a quick check to see that all was clear of the propeller, a cry of 'contact', press the starter button and the engine would come to life with a bang and a puff of black smoke quickly swept away by the slipstream.

Many pilots will recall the preparations for flight but the moment of take-off was also the time of maximum tension which was soon followed by the exhilaration of the feel of enormous power to pull one clear of all those earthbound troubles and into another world. The Spitfire, of all aeroplanes, gave the feeling of freedom of the third dimension. The ability to climb, dive or go in any direction without limit was awe inspiring and diminished man's natural fear of height and gravity. It was this feeling which in retrospect has acquired so much respect and added to the aeroplane's reputation. Those pilots who fought in them will always tell you with a touch of pride, 'I flew Spitfires!' It tells you everything.

Don Robertson

Three clipped wing Spitfire Mk IX/XVIs 'starring' in a film, bearing Normandy D-Day stripes, the codes actually used by No. 340 Sqn on its Spitfires, and Free French emblems.

Index